AF690270

A Trilogy ~ Introduction of the *Theory of Electricity*

What stands out immediately is that this book isn't trying to teach electricity the way textbooks do. It's trying to help readers **finally make sense of it**. Instead of starting with formulas or abstractions, this book starts with curiosity, why things work, where power actually goes, how ideas evolved, and why modern conversations around energy, transportation, and climate often collapse in. The humor isn't decorative, it's structural, it lowers resistance, readers stay with concepts that would otherwise feel inaccessible or intimidating.

This book is ideal for a STEM curriculum by making electricity approachable for all readers, which is exactly where its value lies

Ed Kramer

ed@ejkramer.com

First edition March 31, 2026
ISBN: 979-8-9990378-1-7

Editor: Margaret Henkel

Graphic Design: Ed Kramer

Illustrator:

Noel Red Buffalo

BRADENTON FL.

redbuffalotattoos.com

phatattoo@yahoo.com

Certificate of Registration

This Certificate issued under the seal of the Copyright Office in accordance with title 17, *United States Code,* attests that registration has been made for the work identified below. The information on this certificate has been made a part of the Copyright Office records.

United States Register of Copyrights and Director

Registration Number

TX 9-167-451

Effective Date of Registration:
August 01, 2022

Registration Decision Date:
September 12, 2022

Dedication:

To all my friends, associates, lighting designers, and the countless professionals who keep the world humming, glowing, charging, buzzing, blinking, brewing, scrolling, re-frigerating and reclining.

This book is dedicated to *you* — the linemen, electricians, engineers, tinkerers, and curious minds , who've ever tried to explain what electricity actually *is*... only to be met with blank stares or someone asking *"So... have you ever seen it?"*

Now you have an answer. It zips through wires, and occasionally dances with toasters. Its name is **Eddie Current** — and he's here to help.

May your circuits stay strong, your grounding stay firm, and your stories always be well lit.

~~From one spark to another~~

Introduction of the
~~ Theory Of Electricity ~~

A Trilogy

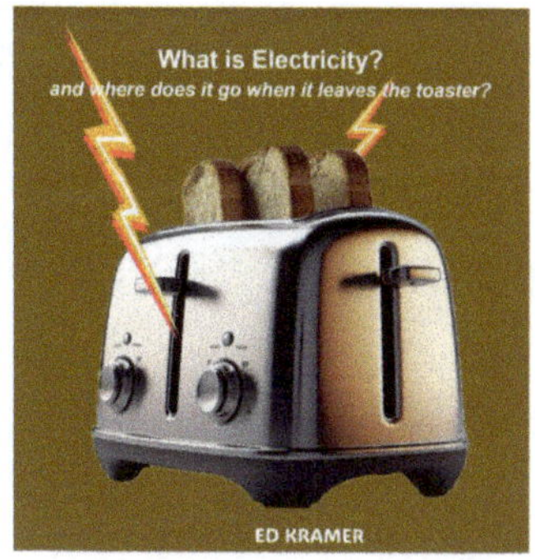

What Is Electricity, and
Where Does It Go When It Leaves the Toaster?

What Is Electricity begins with a question most people have wondered about but rarely ask aloud. This book invites readers to pause, look beyond the wall outlet, and reconsider the everyday miracle they rely on without thinking. Accessible to curious adults and younger readers alike, it establishes the tone of the series: inquisitive, friendly, and gently provocative. It reminds readers that not knowing is not a failure — it's an invitation.

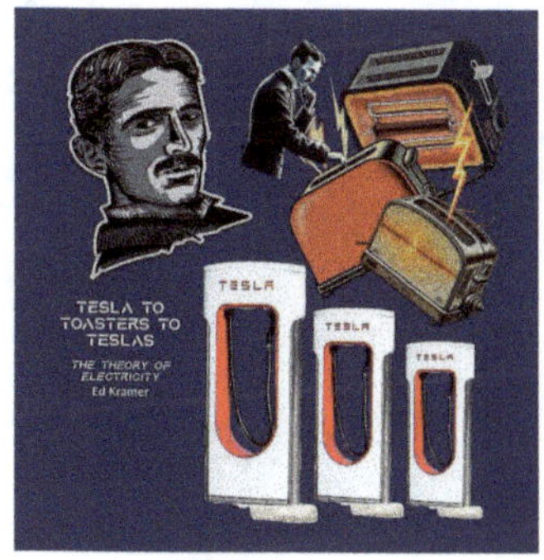

Tesla To Toasters To Teslas
The Theory of Electricity

Tesla To Toasters To Teslas is a clever, accessible, and genuinely original exploration of electricity as both a scientific force and a cultural companion. The book succeeds in doing something rare: it demystifies a complex subject without dumbing it down, using humor, curiosity, and everyday objects to invite readers into the story of power.

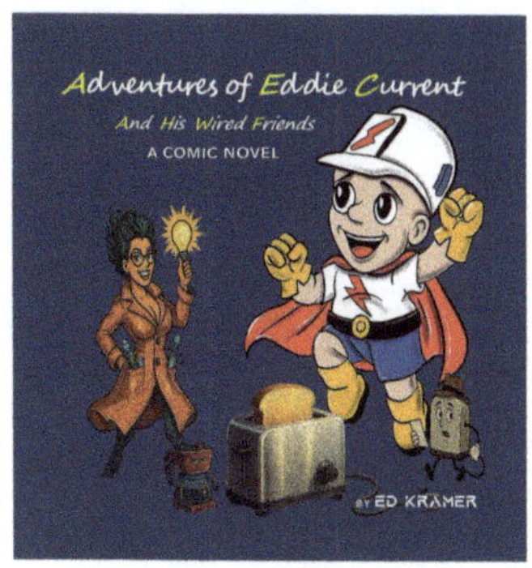

Adventures of Eddie Current
and His Wired Friends

inally it introduces electricity not as a technical subject, but as a character-driven adventure. By personifying electrical concepts, you transform invisible forces into relatable personalities: Eddie Current becomes movement and energy, resistance. Switches and circuits become challenges and choices, Electricity is no longer abstract — it's alive, playful, and purposeful.

These books teach us that technology is understandable, not magical or intimidating.

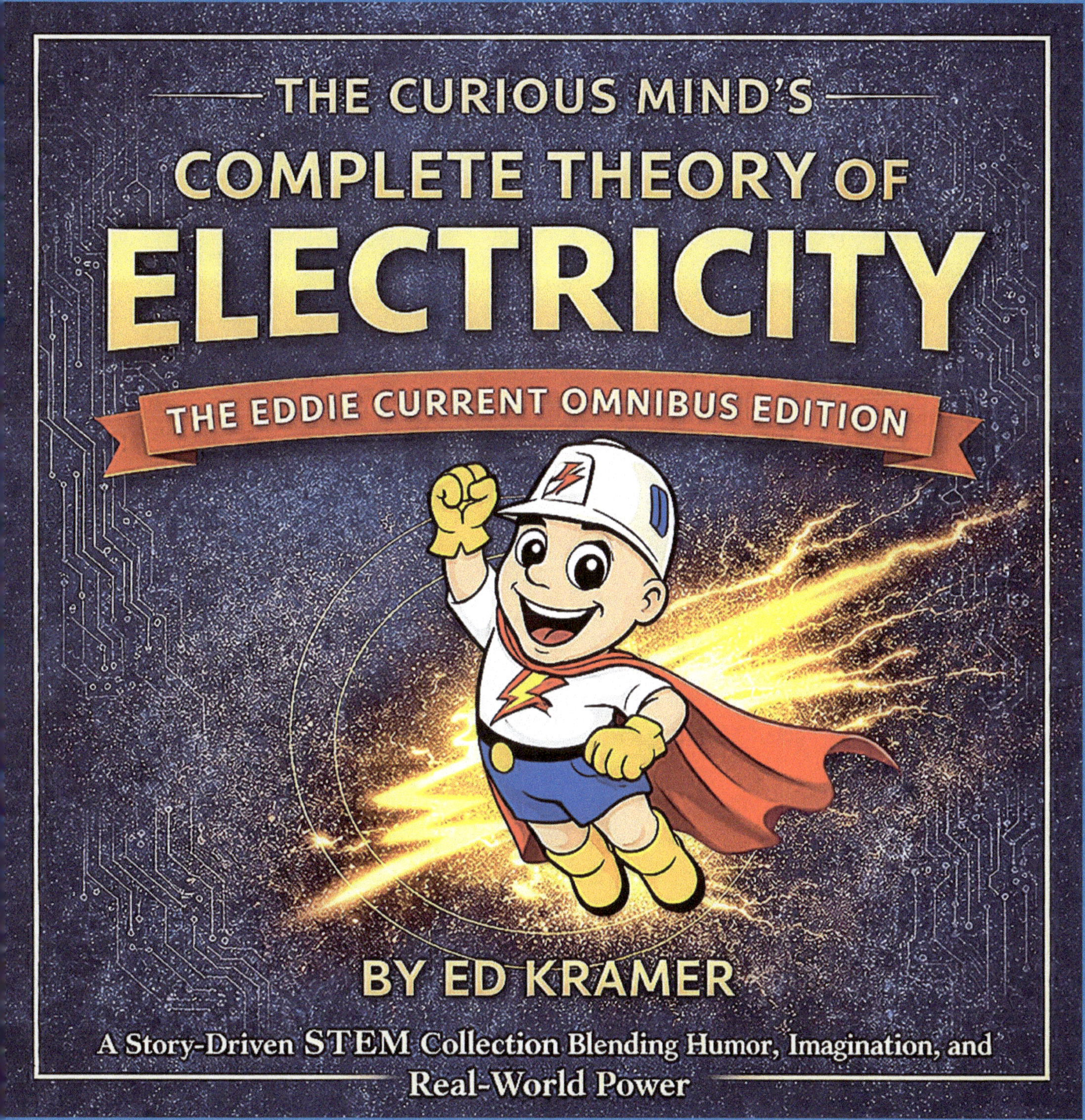
THE CURIOUS MIND'S
COMPLETE THEORY OF
ELECTRICITY
THE EDDIE CURRENT OMNIBUS EDITION
BY ED KRAMER
A Story-Driven STEM Collection Blending Humor, Imagination, and Real-World Power

TABLE OF CONTENTS

TABLE OF CONTENTS cont.

BOOK 1

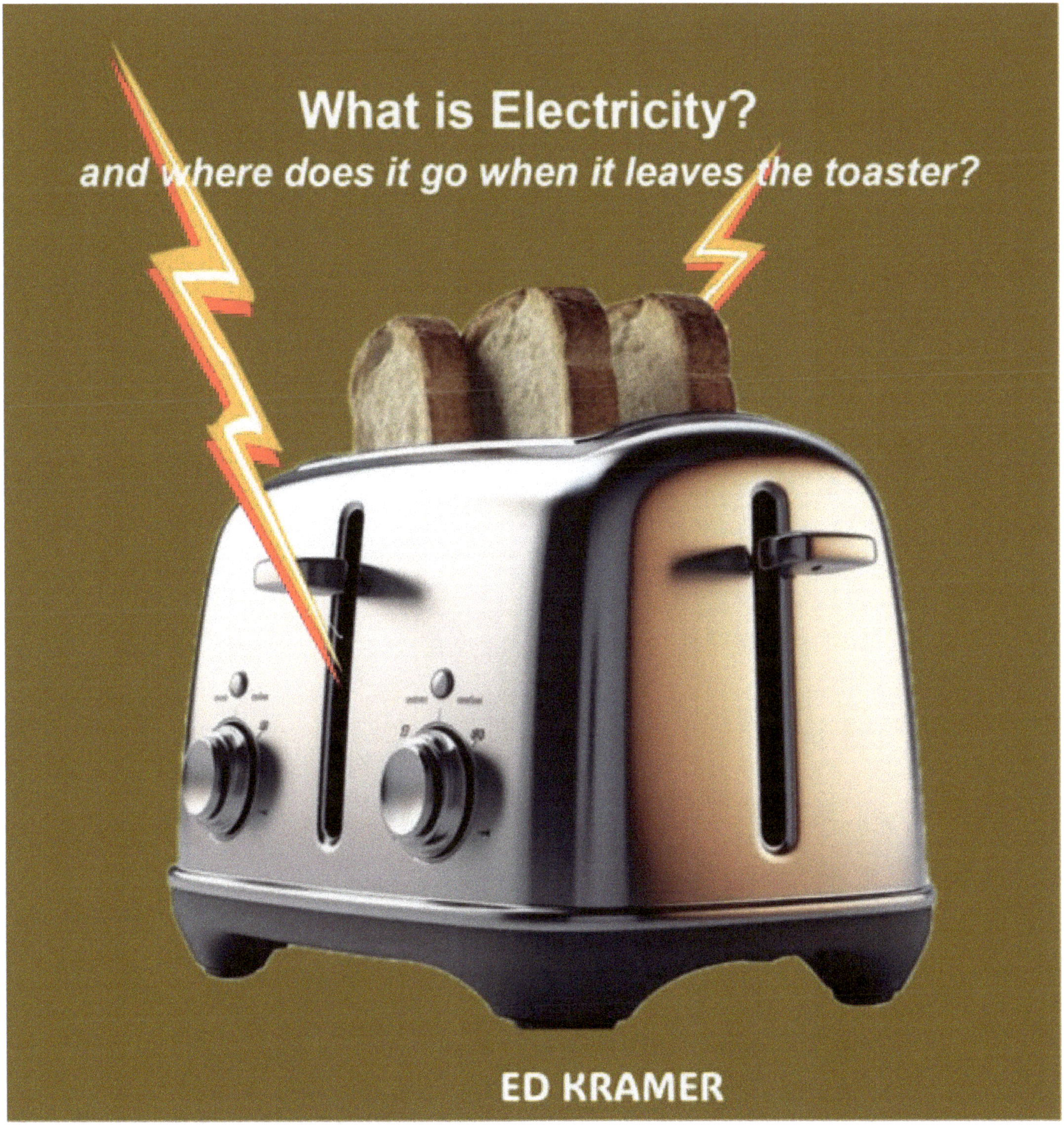
What is Electricity?
and where does it go when it leaves the toaster?
ED KRAMER

Acclaimed by Friends and Business Associates,,,

"I have known Kramer since the early 1990's, and have always enjoyed his wit and friendship. This book is very enjoyable reading."

Bryan H Jones, Manager of Communications, TVA

"Ed always had away of finding humorous ways of looking at challenges and has once again done this with his book."

William Daniels, Marketing Manager, TVA

"If Red Skelton had sold kilo-watt-hours, he would have written this book."

Suzanne Lansford, Consulting Engineer

"Failure to read this book before starting an electrical project is like jumping out of an airplane without a parachute."

H. G. Davis, Business Owner PhD. SOHK

"This book demonstrates a keen understanding of a complex subject, and he is a great tennis player."

Bert Martinez, CPA, Electric Distributor Auditor

" Ed's hysterical (historical) approach to a complex subject is very enlightening and helps us laymen understand the current approach to Green Energy and the Climate Change."

Rick Kauffman, Consulting Engineer

Additional Remarks from Industry Experts,,,

- A great communicator. I have always enjoyed his wit.
- ***Izzy Able, Communications Manager***

- Ed's sense of humor cracks me up. Glad to see that he hasn't changed a bit.
- ***Ray D. Aider, Chief Mechanic, AutoZone***

- The work of a dysfunctional mind.
- ***Hal Apenyo, Manufacturer's Representative***

- Someone will probably sue him.
- ***Dom N. Attion, Consulting Engineer***

- A great depth of understanding of where we are in the world.
- ***Jay L. Bate, Philosopher***

- This lack of understanding could fill a library.
- ***Tim Burr, Lumberman***

- Sooo much more than a Mental Midget.
- ***M. Bezzler, CFO***

- The addition of photos helps us see the BIG PICTURE .
- ***Phil A. Buster, Senator***

Introduction

It was the mid-1990's, and Ronald Regan had just deregulated the electric industry and I was bored.

Introduced 1992

With deregulation, TVA, (Tennessee Valley Authority), which had once monopolized the generation and distribution of electricity, suddenly found itself on an alien playing field, rife with looming competition and in need of a real marketing manager, no, an entire marketing department.

Now they had to woo their 160+ power distributors and directly-served industrial customers in a seven state area, or lose their business. In dire need of a marketing manager, they hired me.

Aged 2021

One early campaign idea I came up with was "***You need it, We got it,"*** and my second brainstorm, ***"Where else can you go?"*** was catchy and to the point, but fell whothe deaf ears of humorless engineers'. Obviously, there was no longer any room for silly campaigns.

The Theory of Electricity was unknown and never discussed except by engineers that tried to impress others with their technical jargon, indeed to explain electricity in terms that common folks would understand.

I created ***"Eddie Current, the Ambassador of POWER,"*** who was a big success, and set off to sell, sell, sell on their clearly stated mission.

But during my windshield time covering the vast TVA territory, my mind wandered as it is often prone to do. This is how the "What is Electricity" story came to be, yes, more than 30 years ago.

There is **NO BULL** in this story because I am an electrical engineer and the editor has a great vision,,,

Everything is "on the level"

Photographed: Lisbon, Portugal & Albuquerque, New Mexico

Before electricity was invented, we relied on candles and created a measure of light known as the **"FOOTCANDLE."** Although this term is still used today, times have changed and "**LEDFOOT CANDLE"** might be a more accurate description,

I created this lightbulb for a 1988 marketing campaign titled "There I$ Money in Lighting."

My daughter insisted that I include this latest scientific theory, and reminded me that it is a great bedtime story to share with your kids.

Darkness Suckers

Electricity in its ultimate form,,,

Initially, it was believed that "light" was emitted by the various devices (light bulbs) that the early pioneers created (more of this story later).

However, the latest scientific data supports another theory.

It turns out that these early devices, as well as the ones we use today, do not emit light at all, but actually ***"suck up"*** the darkness in a room, thereby giving rise to the new scientific theory of *darkness suckers.*

Here is how this works. Eddy Current, who was introduced earlier in this book and is actually captured in this photo inside a lightbulb, goes running through all the electric powerlines in the grid. On the way he drops into houses, scoops up any darkness that has been sucked up by the lightbulbs, and carries it back to the power plant. There it is compressed and shot out of the smoke stack in the form of ***compressed darkness***. (Sounds so logical—it must be a validation)

The process is complicated by the fact that the darkness coming out of smoke stacks would normally be very black, but current government regulation requires that it to be cleaned. For this reason, darkness scrubbers were invented and today we only see clean, white darkness coming out of smoke stacks.

Architects, who understand the need for darkness suckers, call upon their engineers to draw the "suckers" into construction plans. Down the line, contractors generally understand what the architects and engineers have in mind, but they may occasionally ask, "are these suckers in the right place?" (Here the contractors are NOT referring to the Old English slang term "...uckers," whose meaning we will not go into here because this is not an X rated book.)

Note:
At this time, the IES, (Illuminating Engineering Society), is considering commissioning a special panel to investigate and verify this latest astounding discovery.

OK. So now that you've read the introduction, here's the original, complete, unredacted, but enhanced version of the story,

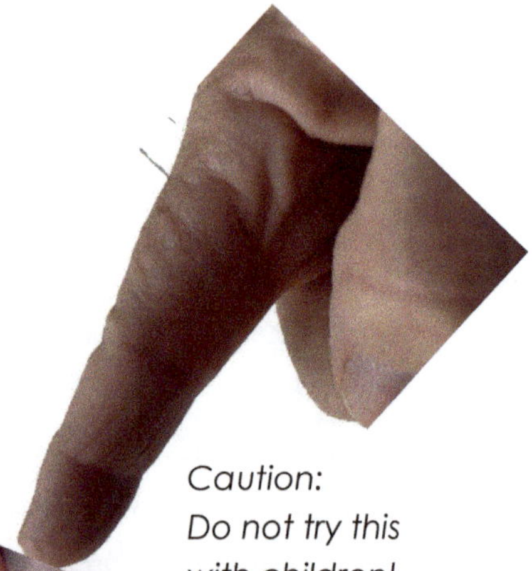

What is Electricity and where does it go when it leaves the toaster?

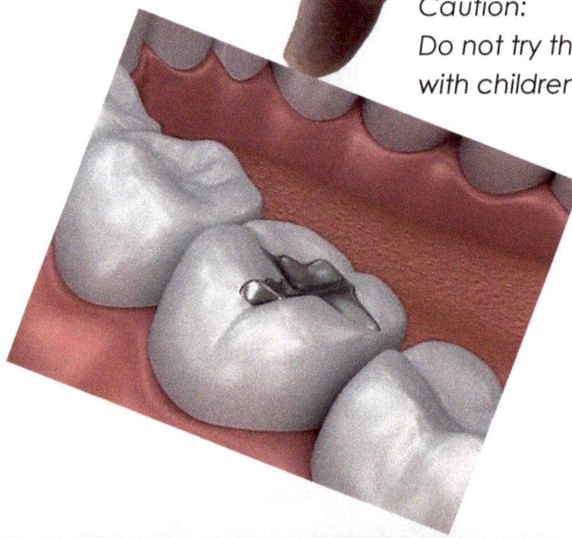

Caution:
Do not try this with children!

Also thugth of as Electricity, first a spark, then the need for society to control a force of nature.

Here is a brilliant experiment that will teach you an important lesson about electricity. On a cool dry day, scuff your feet along a carpet, then reach your hand into a friend's mouth and touch one of her dental fillings.

Here's the scientific question...
Did you notice how your former friend twitched violently and screamed in pain?

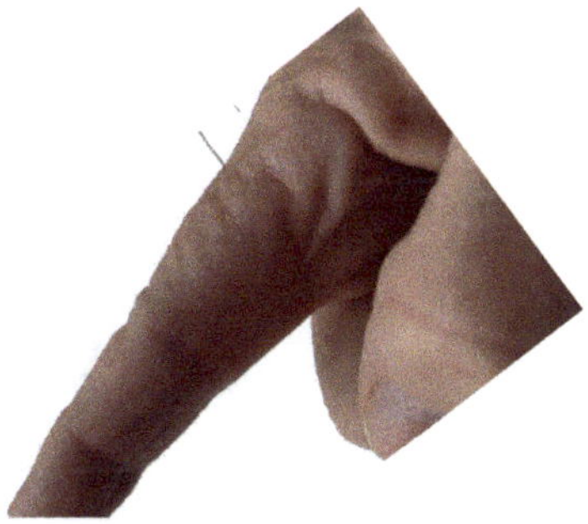

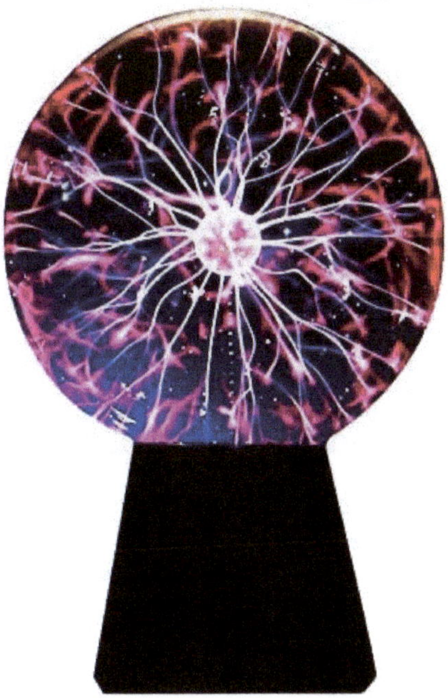

This teaches us that electricity can be a very powerful force, and we must never use it to hurt others, unless we need to teach them an important lesson.

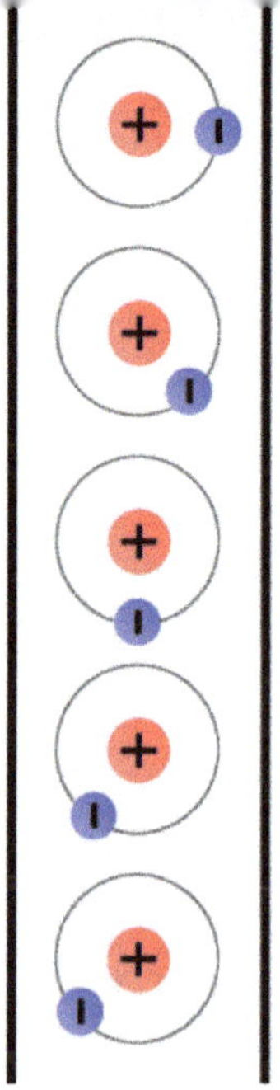

This also illustrates how an electrical circuit works.

When you scuffed your feet, you picked up batches of "electrons," which are very small objects that carpet manufacturers weave into carpeting so that they will attract dirt. The electrons then travel through your bloodstream and collect in your fingertip, where they form a spark that leaps to your friend's filling, then travels down to her feet and back into the carpet, thus completing the circuit.

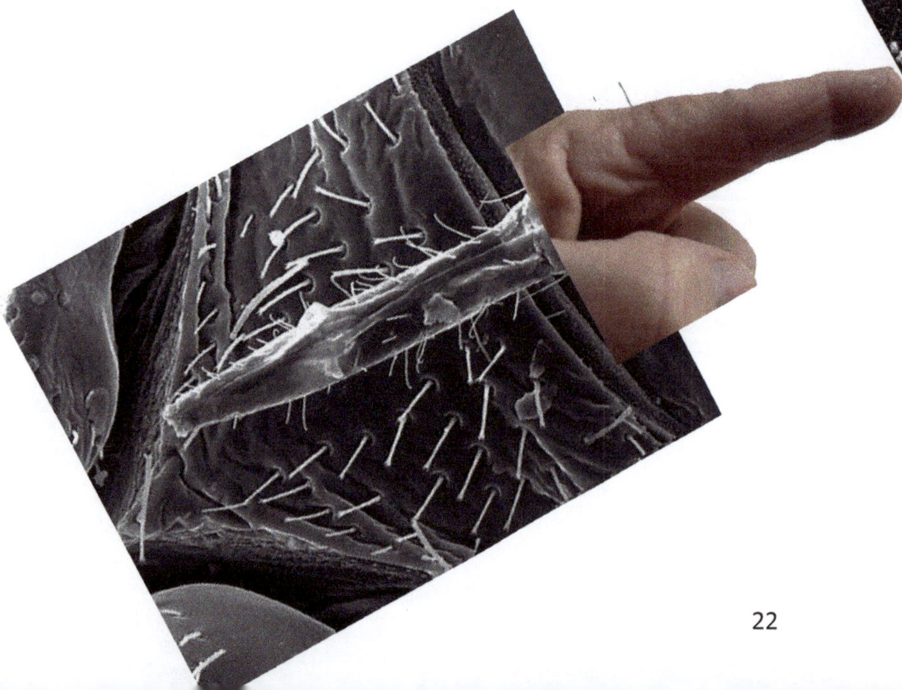

HERE'S AN AMAZING FACT ABOUT ELECTRICITY:

If you scuff your feet long enough without touching
anything, you will build up so many electrons that your finger will explode!
But this is nothing to worry about unless you have carpeting or metal fingers.

Although we modern people tend to take our electric lights, radios, mixers, etc. for granted, hundreds of years ago people did not have any of these things, which is just as well because there was no place to plug them in.

Where is the toaster??
How do you expect me to
warm my Pop-Tarts???

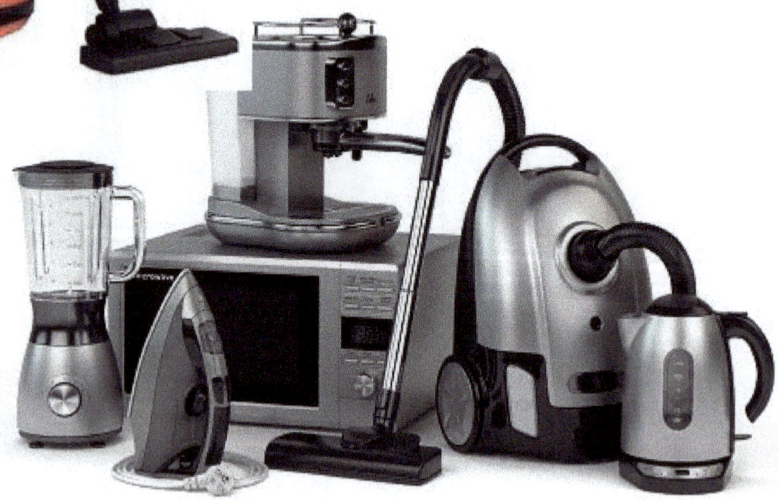

Then along came the first Electrical Pioneer, Benjamin Franklin, who flew a kite in a lightning storm and received a serious electric shock.

Photographed: Ben Franklin & Money

This not only proved that lightning was powered by the same force as carpets, but it also damaged Franklin's brain so severely that he began to speak in proverbs, such as, "A penny saved is a penny earned." Eventually he became famous, accumulated many $100 dollar bills and was given a job running the post office.

Is this "THE GRID?"

Photographed:
Florida & Tennessee

After Franklin, came a stampede of Electrical Pioneers whose names have become part of our electrical terminology: Myron Volt, Mary Louise Amp, James Watt, Bob Transformer, Charlie Capacitor, etc.

These pioneers conducted many important electrical experiments and became famous in their own right.

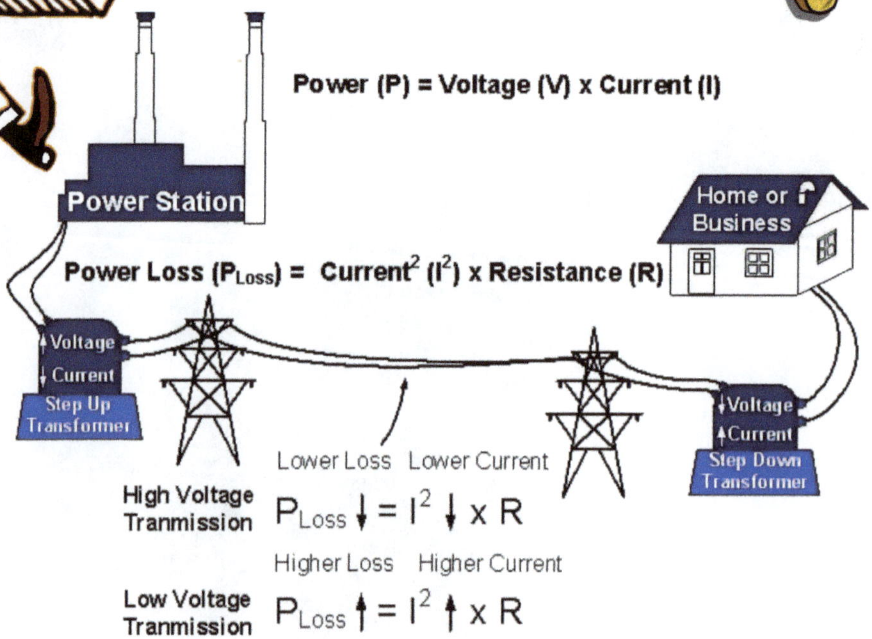

Another pioneer we never hear about is Charles Fritts: (1850-1903) , a U S inventor who pioneered the development of the SOLAR PANEL

Photographed: Florida

For example, (this is true), in 1780, Luigi Galvani discovered that when he attached two different kinds of metal to the leg of a frog, an electrical current developed and the frog's leg kicked, even though it was no longer attached to the frog, which was dead anyway. Galvani's discovery led to enormous advances in the field of amphibian medicine. Today, skilled veterinary surgeons can take a frog that has been seriously injured or killed, implant pieces of metal in its muscles, and watch it hop into the pond like a normal frog, except for the fact that it sinks like a stone. Also, since water is an efficient conductor of electricity, the frog is immediately electrocuted.

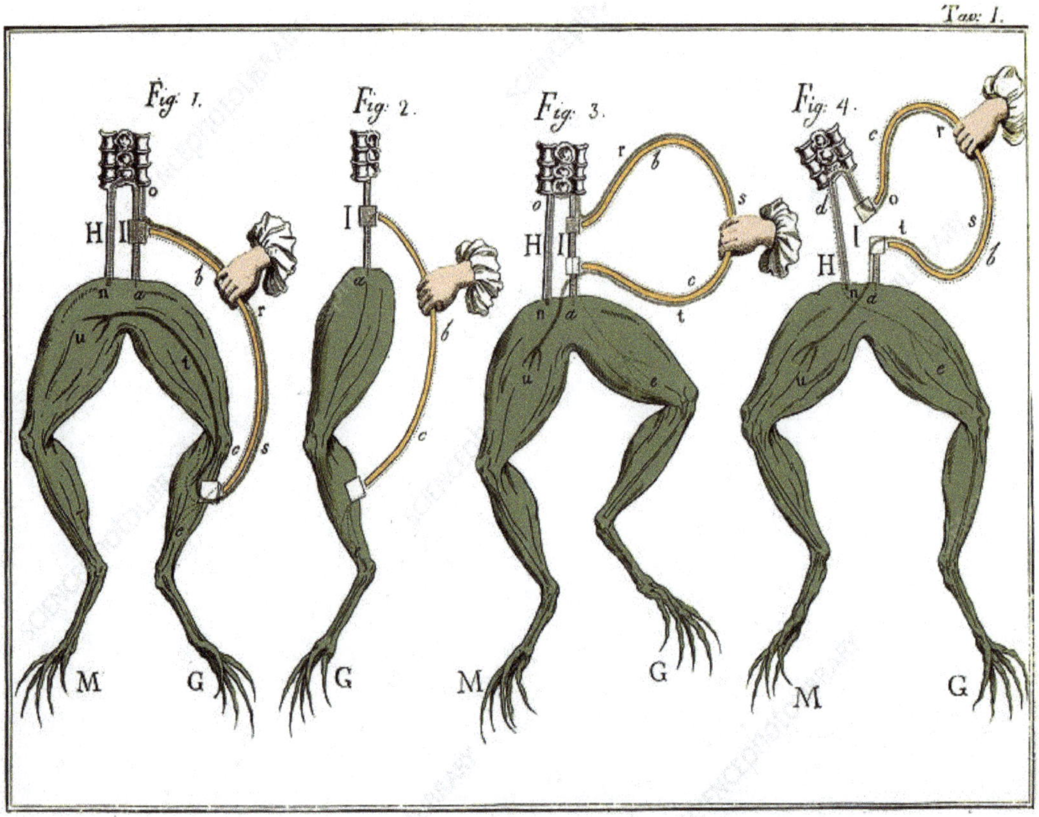

The greatest Electrical Pioneer of them all was Thomas Edison. He was a brilliant inventor despite the fact that he had little formal education, lived in New Jersey and vacationed in Florida to get away from the snow.

Edison's first major invention, in 1877, was the phonograph, which could soon be found in thousands of American homes, where it basically sat until 1923, when the record was invented.

Photographed:
Chattanooga, Tennessee
& New Delhi, India

But Edison's greatest achievement came in 1879 when he invented the electric company, which is functionally just a brilliant adaptation of the simple electrical circuit. The electric company sends electricity through a wire to a customer...
... then immediately gets the same electricity back through another wire, then...

(this is the brilliant part)

...sends it right back to the same customer again!

So now you see how electric companies can sell their customers the same batch of electricity over and over, thousands of times a day. They never got caught, because very few customers take the time to catch and examine their electricity closely. The last year any new electricity was generated was in 1937,,,

1992

2021

Electric distribution system photographed: Southeast Asia

Is Solar making inroads,,, or just giving the illusion of Going Green

How many volts,
coming or going,
240 OR 241?

Electric distribution system photographed: India & Florida

Is this an electrical evolution or is it an electrical revolution?

SHOCKING!

Electric distribution system photographed: India

All electric companies, regardless of their source of energy, have been re-selling the same electricity for decades, which is why they have so much free time to apply for rate increases.

Photographed: Nuclear Power Plant, Ohio

By the way,,,
How many power plants have you had the opportunity to explore?

For instance, on this wind farm, the electrons may have originally came from carpeting!

Here is another example of a distribution system designed specifically to keep customers confused.

Photographed: Electric distribution system Southeast Asia

This is the modern method of creating Alternating Current (AC).

Photographed: Electric distribution system Southeast Asia

Lets put a spin on this story,,,

Thanks to my TVA training, I was able to stop some of the smoke and pollution. It's a highly technical government thing, so don't try this at home.

Photographed: Ohio & Florida

Living up to the reputation of EDDIE CURRENT,
I spin the blades very fast!

I wish I were renewable too!

Today, thanks to men like
Edison and Franklin, and frogs like
Galvani's, we receive
almost unlimited benefits
from electricity.

For example, in the past decade scientists have developed the laser, an electronic appliance so powerful that it can vaporize a bulldozer 2000 yards away, yet so precise that doctors can use it to perform delicate operations to the human eyeball, provided they remember to change the power setting from "Vaporize Bulldozer" to "Delicate."

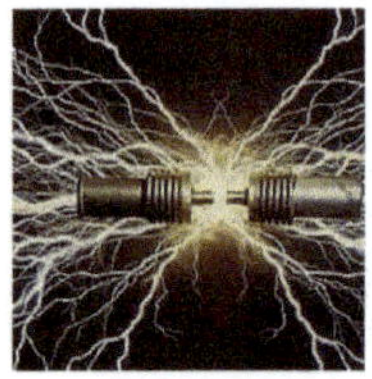

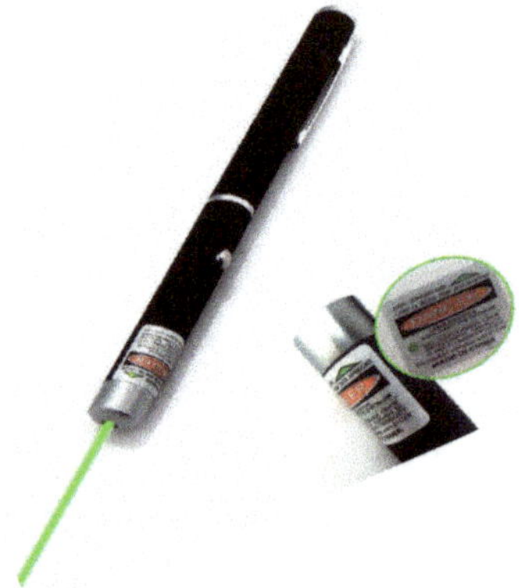

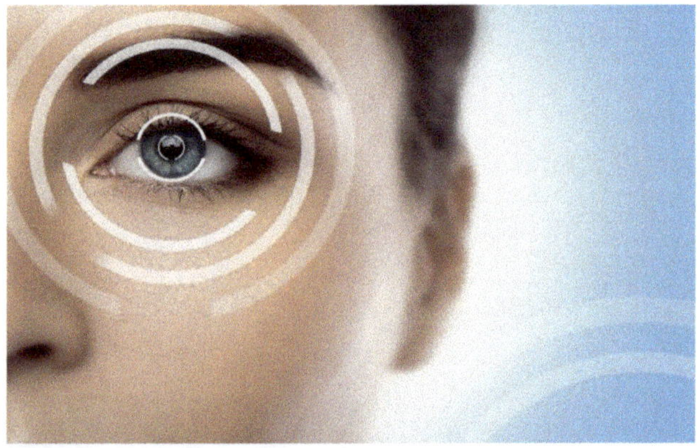

Photographed: Florida

THE END!

FOR HISTORICAL TRANSPARANCY,
HERE IS THE STORY OF ELECTRICITY IN ITS ORIGINAL FORM,,,

Today's question: What in the world is electricity and where does it go after it leaves the toaster?

Here is a simple experiment that will teach you an important electrical lesson: On a cool dry day, scuff your feet along a carpet, then reach your hand into a friend's mouth and touch one of his dental fillings. Did you notice how your friend twitched violently and screamed in agony? This teaches one that electricity can be a very powerful force, but we must never use it to hurt others unless they need to learn an important lesson about electricity.

It also illustrates how an electrical circuit works. When you scuffed your feet, you picked up batches of "electrons," which are very small objects that carpet manufacturers weave into carpet so that they will attract dirt. The electrons travel through your bloodstream and collect in your finger, where they form a spark that leaps to your friend's filling, then travel down to his feet and back into the carpet, thus completing the circuit.

AMAZING ELECTRONIC FACT:
If you scuffed your feet long enough without touching anything, you would build up so many electrons that your finger would explode! But this is nothing to worry about unless you have carpeting.

Although we modern persons tends to take our electric lights, radios, mixers, etc. for granted, hundreds of years ago people did not have any of these things, which is just as well because there was no place to plug them in. Then along came the first Electrical Pioneer, Benjamin Franklin, who flew a kite in a lightning storm and received a serious electrical shock. This proved that lightning was powered by the same force as carpets, but it also damaged Franklin's brain so severely that he started speaking only in incomprehensible maxims, such as, "A penny saved is penny earned." Eventually, he had to be given a job running the post office. After Franklin, came a herd of Electrical Pioneers whose names have become part of our electrical terminology: Myron Volt, Mary Louise Amp, James Watt, Bob Transformer, etc.

These pioneers conducted many important electrical experiments. For example, in 1780 Luigi Galvani discovered (this is the truth) that when he attached two different kinds of metal to the leg of a frog, and electrical current developed and the frog's leg kicked, even though it was no longer actually attached to the frog, which was dead anyway. Galvani's discovery led to enormous advances in the field of amphibian medicine. Today, skilled veterinary surgeons can take a frog that has been seriously injured of killed, implant pieces of metal in its muscles, and watch in hop back into the pond just like a normal frog, except for the fact that it sinks like a stone.

.

But the greatest Electrical Pioneer of them all was Thomas Edison, who was a brilliant inventor despite the fact that he had little formal education and lived in New Jersey. Edison's first major invention in 1877 was the phonograph, which could soon be found in thousands of American homes, where it basically sat until 1923, when the record was invented.

Today we believe Edison's greatest achievement came in 1879 when he invented the electric company. Edison's design was a brilliant adaptation of the simple electrical circuit: the electric company sends electricity through a wire to a customer, then immediately gets the electricity back through another wire, then (this is the brilliant part) sends it right back to the customer again.

This means that an electric company can sell a customer the same electricity over and over, thousands of times a day and never get caught, since very few customers take the time to catch and examine their electricity closely. In fact, the last year any new electricity was generated was in 1937.

Today, thanks to men like Edison and Franklin, and frogs like Galvani's, we receive almost unlimited benefits from electricity. For example, decade scientists have developed the laser, an electronic appliance so powerful that it can vaporize a bulldozer 2000 yards away, yet so precise that doctors can use it to perform delicate operations to the human eyeball, provided they remember to change the power setting from "Bulldozer" to "Delicate."

THE END!

This page is left blank for your notes:

BOOK 2

LET THE SPARKS FLY!

TESLA TO TOASTERS TO TESLAS

THE THEORY OF ELECTRICITY

Ed Kramer

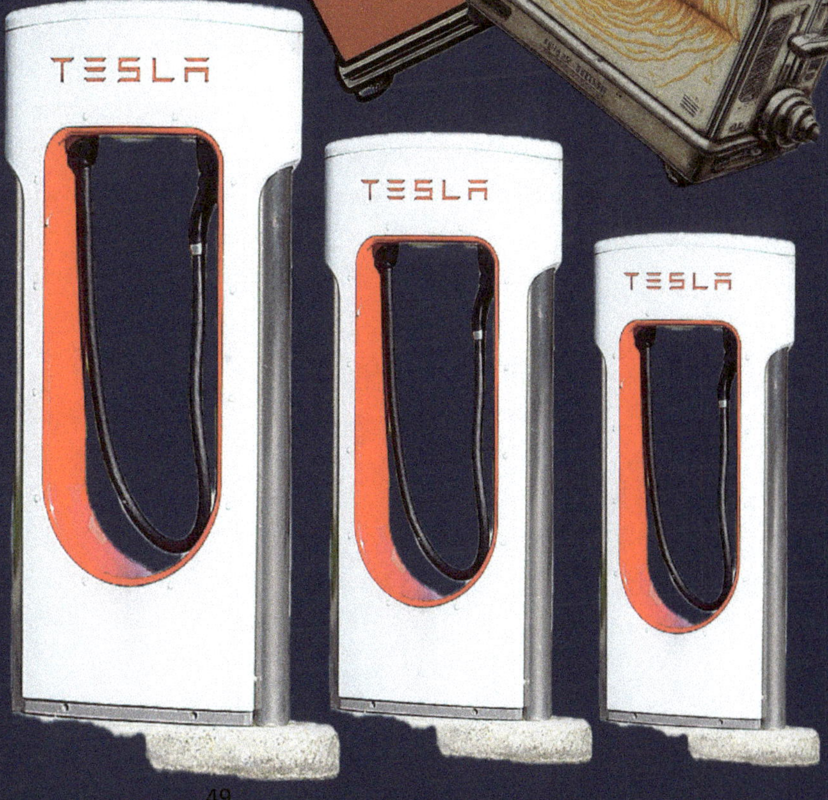

THEORIES OF ELECTRICITY

T O E = *T*heory *O*f *E*lectricity or *T*heory *O*f *E*verything

**This book has a great amount authenticated detail,
awesome wit and factual references, a lot of photos, and with portions written
to question the prevailing thoughts commonly found in the press,
taking a deep dive into the absurdity of many common beliefs.**

As a matter of national security — I have to share this total story!

Eddie Current

"Stop that immediately"
shouted Kramer, Senior as he raced across the room toward me. Sparks shot out of the receptacle and the lights went out in the entire house. You guessed it, my interest and infatuation with electricity began at a very early age. This was partially encouraged by my father who was an electrical engineer working for a municipal electric power company.

As I grew up, birthday and Christmas gifts at an early age involved electrical products, but focused on those that had educational value. Many of the gifts came from a company called HealthKit. At the time, they made the world's finest kits for radios, amplifiers, and shortwave, where you could build professional quality electronics equipment yourself, even if you had little or no experience. As my father raced across the room, the lights went out, he stumbled on a chair and swore to teach me a lesson I would never forget.

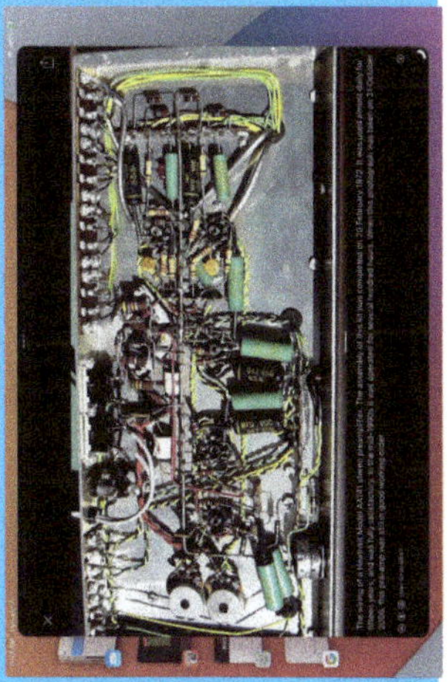

My knowledge and interest in electricity took off like a rocket headed to Mars.

In the first chapter, you will meet Eddie Current, who will be your guide thorough this complex maze. starting with the Theory Of Electricity, and then on to the challenges of electricity and Climate Change.

The purpose of this book is to have some fun, laugh at some the absurdities of the climatologists and government, and to gain a bit of knowledge about our powerful energy sources and uses.

Lets start by introducing
Eddie Current.
Growing up is a tough situation for electrons.

Early childhood toys,,,

Eddie Current at an early age of about 8 told his dad, Mr. Kilowatt, that he wanted to be a pirate when he grew up.

Dad was quite surprised 'cause most young electrons at that age wanted to be cowboys or firemen.

Thank goodness dad did not take him seriously and set a schedule for peg leg surgery and an eye removal!

As he grew up, he found enjoyment in comic books, mostly those involving Super Heroes.

Eddie Current's favorite flavor of ice cream is Shock-a-lot

But becoming a **Super Hero** was not his calling. "If I could become one, I would have to have the power of empathy. Empathy allows us to understand and feel the experiences of others, which empowers us to take action and make a positive difference in someone's life." He imagined that if he all had the power of empathy, he would be able to truly listen to and understand without judgment or bias. He could use his power to break down barriers and create connections between people of different backgrounds, cultures, and beliefs. "We could use it to promote kindness, understanding, and compassion throughout the world."

This seemed to be too limiting to him, and his skills and ambitions were maturing.

On a trip to Las Vegas he experienced the Sphere of Influence,,, and was inspired. This book is Eddie Current's Sphere of Influence. Eddie Current found his calling and was appointed by the government as Ambassador of Power.

So let's tap into our own superpowers and embrace empathy. Let's use our unique abilities to make a difference in the lives of those around us and work towards creating a more empathetic and connected world. Together, we can make a powerful impact on the world and create a brighter future for all.

If you wanted to become a motivational speaker or writer, I believe that each and every one of us has the potential to make a positive impact on the world, no matter how big or small. And while superpowers may seem like something out of comic books, they actually serve as a great metaphor for the power we all possess within ourselves.

WITH GREAT POWER COMES GREAT RESPONSIBILITY,,,

"If I saw someone harassing some electricity, would that be an abuse of power."

Understanding Electricity is complex and most readers find it hard to follow all the details. But the underlying ideas can be clearer than the details.

Information is an important tool, but with a bit of humor interjected, the reading goes much faster. So, my goal to give you an explanation of some of the tasks that are critically dependent on electricity as well as an explanation of many of our every day encounters with this mystery of this magic.

Where to start? Some folks suspect that energy like money flows in and out of your life, and others suspect that it only flows out. Today, energy and its generation are major conversation points in every discussion when the topic gets to climate control. We often hear the terms green energy, climate change, humans are the main cause of climate change and the creation of green house gasses.

Hopefully, we will demonstrate the interrelationship of these commonly used terms. Understanding energy in this book is focused exclusively on electricity.

The first book describing electricity was titled *"What Is Electricity,,, And Where Does It Go When It Leaves The Toaster"*. You met ***Eddie Current***, *Ambassador of Power.* Eddie awoke, and like with Rip van Winkle, who wakes 30 years later, saw the whole scene of electrical power and its distribution had changed. These chapters help bring us up to speed starting with Tesla, to Toasters, and back to Teslas.

Are you ready for a thorough explanation of the legends and the misinformation that has inundated our lives and our dependency on this miracle?

We are going to squeeze as much as possible out of Eddie Current, so fasten your seat-belt and start flipping the pages.

Lets clear the air. If you are a legalist, then you might think that this is an eddy current.

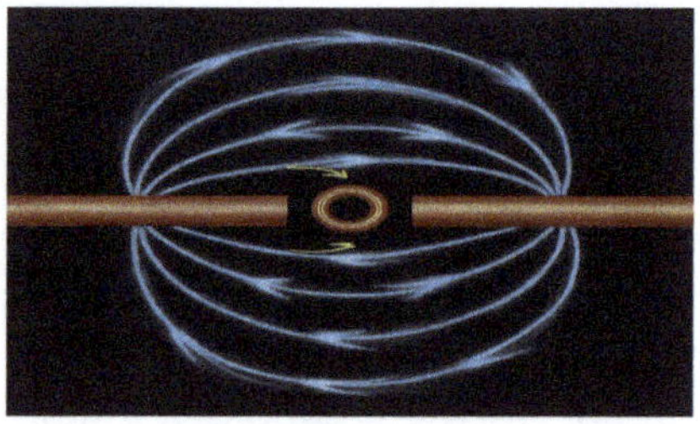

But it is not! He is actually a multi-skilled fella doing the many tasks we require of him. Sucking up the darkness in our homes, powering our TVs and music systems and laptops, charging our smart phones, cooking our breakfasts lunches and dinners, powering our EV automobiles, and we are just getting started.

On the following pages, you will discover several of the Eddie Current specialists. Maybe they are doctors. Some are internists, others might specialize in specific fields, like neurosurgery, pulmonary, orthopedics, pediatrics or gastroenterology. Do you think it is the same electrons that turn the wheels of your electric vehicle which also powers your smart phone? Do you think that they all come from the same source? Are the electrons that come from coal and oil power plants the same as the ones that come from the wind. What about the Eddie Currents that come from solar panels? What about the Eddie Currents that come with the ever-ready batteries you buy in the store, or the lithium batteries you have in your smart phone?

Maybe electrons look like a glass of orange juice, or a bottle of Mountain Dew. I am confident we all have referred to electron flow as turning on the juice!

What is an Eddie Current? He's not a person. He's not an Animal. He doesn't have a gender or a nationality. No one knows what he looks like. What is a kilowatt, or a kilowatt-hour? Can you get a bucket full at the supermarket, or on Amazon? What ia an amp, or a volt? Eddie Current will help guide you through these complex subjects.

In the beginning, Nichola Tesla called all electrons before he knew they had their own personalities. Now, you have the opportunity to learn all about them and will meet their leader, Eddie Current.

Maybe he is a Super-Hero, invisible, but doing all the things that we are dependent on. But where do all these Eddie Currents come from? Where are they stored? Where do they live?

Let me think,,, It seems that in almost every discussion about electricity the word 'Ground' comes up. I know, this only comes up when talking about how to hook up an appliance, or wire a house, but it is a common term. Like you need the black wire hooked up to the 'Hot' and the white wire hooked up to the 'Ground.' Or when a fuse is blown, another term is 'short circuit,' which probably means how long does it take the electrons to get to 'Ground.

Continuing this train of thought, electrical inspectors regularly check if the installation is properly Grounded.

This may be a hard concept to visualize. I'm sure you have probably dug in the garden or planted flowers around the house, and in all your digging, I'll bet you never uncovered an electron!

Does Eddie Current have the ability to travel invisibly or is HE the *FORCE!*

If you hear someone referring to electricity as "Juice," please demonstrate your expertise in the subject and inform them that Juice is a meaningless colloquial slang term. It should not be used in any serious discussion about electricity.

Electricity has to do with the presence and movement of electrons. Electricity is not power, it is a "carrier" of power.

Let's start with a few explanations,,,

Eddie Current often travels into the ground because it is an easy path and there he finds a home where he is welcomed, loved, given food and a place to recharge.

The major reason that all homes have a 'Ground Wire' is to give Eddie Current a two-lane highway to his favorite restaurant.

A sweater I bought was picking up static electricity. So, I returned it and they gave me another one free of charge."

Electrical Terms: Faults vs Shorts

Electric faults are not a blameless crime. Even the geologists who look for faults could be clueless!

On the other hand, electric shorts are caused by electrons eagerly trying to get back to their home in the ground. If you are where lots of trees were downed by a hurricane, do you have a short or a fault? You can see the cause, therefore, this is a short.

For the clarification and an actual definition, we asked Eddie Current to weigh-in:

Here is a definition you can take to the bank:
A short circuit has zero resistance between two Wires / Circuits / Systems, on the other hand in a Fault, current has a resistance that draws current. The amount of resistance decides how much current is drawn and could be caused by a breakdown in the insulation of a system...
Are you now ready to explain this to your friends. Probably more than you wanted to know, now, whose fault is this???

The Power Surge:

Lets examine another **Electrical Term**.

A power surge is when Eddie Current gets aggravated by the false theories are trying to take hold. His adrenalin surges and the elevated levels of power will generally cause TVs, smart phones and iPads to turn off, or, in technical terms, cause a non-passive failure. Now that he has your attention, he then pauses for a few moments to let the climate activists cool off.

Examples of common household "Surge Protectors". Consult Sparky, the Electrician for installation,

CLIMATE CHANGE

WE CAN'T FIGHT
HOMELESSNESS, HUNGER, OR POVERTY
BUT WE ARE GOING TO FIGHT "CLIMATE CHANGE"
JUST LET THAT SINK IN FOR A MINUTE.

Another load of
electric car fuel
headed to the
power plant.

What is greenhouse gas?

A day does not go by without you hearing the term 'Climate Change' uttered, usually as the blanket explanation for all of our wows. From natural disasters to man-made catastrophes, the news media to social media to casual conversations all blame 'Climate Change', with the implied and often stated solution of government mandated regulations.

There's a full-blown crisis, mandating conversion to green. The government is mandating, conversion to green without considering the cost, and the dictation of this conversion is eliminating free choice of American citizens. Vehicles emit carbon dioxide, the most common human-caused greenhouse gas. What is greenhouse gas? Surprise! Its mostly carbon dioxide... But most people realize that carbon dioxide is actually a benefit to the planet, so they had to change the name to greenhouse gas. Its liked a full-blown imitation of Howard Hughes going totally off radar.

We are inundated with statements like: California's just-announced plan to phase out sales on new combustion engine vehicles by 2035. This is doomed to fail. "How can the state electrify the vehicle fleet if it can barely keep the lights on?"

California recently asked homeowners to reduce electricity consumption to help avoid blackouts as temperatures soared and the power system struggled to keep up. The plea was effective, with consumers temporarily dialing back demand enough to keep the lights on across the state. But these sorts of close calls are the stuff of nightmares for system operators, and this specific brush with near-disaster had a new element that caught a lot of attention: a call to electric vehicle owners to avoid charging during peak demand hours.

These types of discussions get emotional quickly, so it's worth stepping back a bit to look at the data. The world is at a turning point where it must switch from fossil fuels to renewable energy sources in order to tackle climate change. Because they are clean, plentiful, and sustainable, renewable energy sources like wind and solar energy are gaining popularity. Yet, one of the primary issues with renewable energy is the intermittent nature of these sources and these energy storage options.

Solar energy systems can only generate electricity during the day, while wind turbines can only do so when there is enough wind. Because of this, it is challenging to rely on these sources for a regular and reliable energy supply.

Thankfully, technology has improved and a variety of renewable storage options are now becoming more readily available. We shall examine the various forms of renewable energy storage on later pages.

https://www.bloomberg.com/news/articles/2022-09-07/a-text-alert-may-have-saved-california-from-power-blackouts

Please note,,,
This is where we all learn to put on socks
and tie our shoes,,

We all long for a bright sunny day...

Jonny Nash's 1972 hit, ***I Can See Clearly Now***

I can see clearly now the rain is gone
I can see all obstacles in my way
Gone are the dark clouds that had me blind

It's gonna be a bright (bright)
Bright (bright) sunshiny day
It's gonna be a bright (bright)
Bright (bright) sunshiny day

Oh, yes I can make it now the pain is gone
All of the bad feelings have disappeared
Here is that rainbow I've been praying for

It's gonna be a bright (bright)
Bright (bright) sunshiny day

look all around, there's nothing but blue skies
Look straight ahead, there's nothing but blue skies

I can see clearly now the rain is gone
I can see all obstacles in my way
Here is that rainbow I've been praying for

It's gonna be a bright (bright)
Bright (bright) sunshiny day

Bri-ri-ri-ri', bright, (bright)
Bright (bright) sunshiny day, yeah, eh

let's make some electricity, yeah,, yeah,,, yeah,

Eddie Current is not speaking out against prevailing public opinion. He is not denying that Climate Change is taking place, but it is obvious that electricity is at the center of any discussion of climate change. This is just a small opportunity to give a voice to other causes and alternatives. With a degree of humor, many of the climate change contributors will be exposed and examined, and you will have the opportunity to form you own opinions.

There is actual data that shows that there is climate change. He is only trying to show that the climatologists may be incorrect in that their fundamental assumptions for the causes are too narrow and the proposed solutions may not be founded in logical reason.

Much of their argument is merely a speed bump the highway in attempting to find a solution. When you apply sound scientific and engineering principals to global climate change, you can avoid much of the road kill along the highway to a better environment.

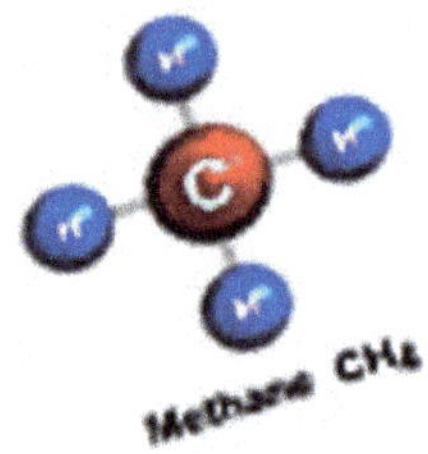

VS

When causes and benefits of corrections are no longer driven by a fundamental assumptions for anthropogenic causes (caused by human activity), solutions will become driven by the desire for an improved global world and will not require government issued incentives. This discussion also validates an assertion that government funding of academic research will also get corrupted by allowing a similar orthodoxy.

Another interesting statement is on the necessity for STEM fields (Science, Technology, Engineering, and Math) to conform to certain dogmas, like anthropogenic climate change. Academics must also change.

Are we living in a POW camp? Or getting our guidance and leadership from a very badly dressed old man?

If you are not a climatologist, prevailing orthodoxy is ostracism comunication from an existing climatology society.

Greenhouse Gases Emitted by Volcanoes, Vehicles, Fossil Fuel Power Plants, Farm Animals, & Rocket Launches

Where does all this CO_2 come from??? A partial list of the culprits include Greenhouse Gases Emitted by Volcanoes, Vehicles, Fossil Fuel Power Plants, Farm Animals, & Rocket Launches.
Which headline will grab your attention? And we all know that the news we are exposed to is driven by advertising dollars and viewer ratings.

Lets start with volcanos. Volcanoes release large amounts of greenhouse gases such as water vapor and carbon dioxide. There have been times during Earth history when intense volcanism has significantly increased the amount of carbon dioxide in the atmosphere and caused global warming.

Today, there are more than 1500 active volcanoes on Earth, with 40-50 continually erupting. Generally around 20 will be aggressively erupting on any particular day. Once the news media has devoted their maximum 90 seconds to the tragedy, they are on the next flash to keep the viewers attention. We never hear about pollution from farm animals, or jumbo jets!

Thinking back to my formative years, I remember a science class in high school. Our chemistry teacher, Louis Pasture, gave us the experiment for the day. We learned that when mixing baking soda and vinegar, it creates a gas called carbon dioxide.
Maybe all the high school science teachers globally are the major culprits in the creation of all this CO_2!

Do you think we should start a fundraiser to send a solar-powered flashlight to every person who still believes the earth is flat.

(I hope they endorse this book)

The FLAT EARTH SOCIETY

The most powerful governments on earth can't stop a virus from spreading... but they say they can change the earth's temperature if you pay more taxes?

Causes and Solutions:

Greenhouse Gasses:

While it is tempting to blame the media for oversimplifying complicated scientific ideas and presenting only the bad news, we must remember that they are catering to the desires of their readers and viewers.

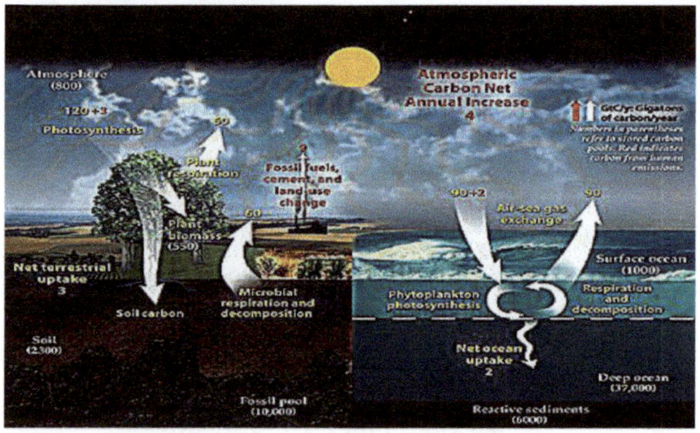

Lets take a look at Greenhouse gasses.

What is CO_2

Carbon dioxide is an important greenhouse gas that helps to trap heat in our atmosphere. Without it, our planet would be inhospitably cold. CO_2 is one of the foundations of our world.

For eons, the world's oceans have been sucking carbon dioxide out of the atmosphere and releasing it again. The ocean takes up carbon dioxide through photosynthesis by plant-like organisms phytoplankton), as well as by simple chemistry: carbon dioxide dissolves in water.

Let's explore this narrative about climate change. Common and often used terminology includes the term "Green House gasses." The major gas in this vehicle emission is —- you guessed it—CO_2, a necessity for sustaining life on this planet. Trees convert CO_2 to carbon and oxygen. The oceans need oxygen, fish need oxygen, and with more oxygen, we get more and bigger fish, bigger trees. In fact everything is much better, bigger and healthier.

Another "Green House gas is Methane CH_4

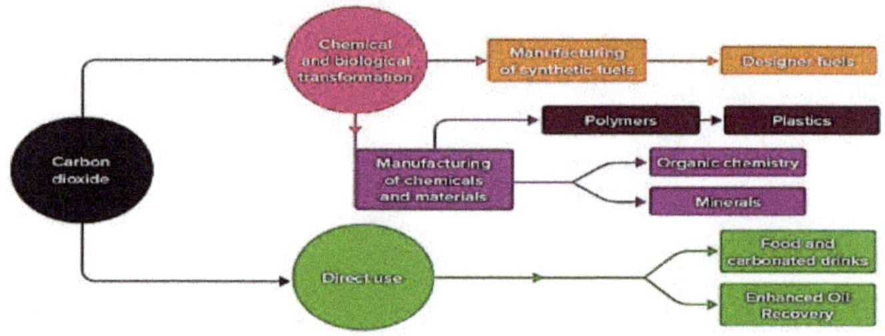

Its colorless, odorless and invisible to the naked eye. Methane (a complex hydrocarbon) is another of the foundations of our world, has important implications for climate change, particularly in the near term.

Two key characteristics determine the impact of different greenhouse gases on the climate: the length of time they remain in the atmosphere and their ability to absorb energy. Methane has a shorter atmospheric lifetime than CO_2 , but it is a much more potent greenhouse gas, absorbing much more energy while it exists in the atmosphere, especially the stratosphere..

CH_4 is responsible for more than 25 per cent of the global warming we are experiencing today. Due to its structure, methane traps more heat in the atmosphere per molecule than carbon dioxide (CO_2), making it **80 times more harmful than CO_2** for 20 years after it is released.

Methane is also a great energy source. Remember when we were being told that we would run out of gas, well, now we have found a way to harvest vast amounts of this precious energy source.

Methane comes from many sources. It exists far below the surface and well below the possible existence of any fossils from prehistoric dinosaurs or trees. In fact, it is probably the source of all the shale oil that we are retrieving today.

We haven't been told the truth about where hydrocarbons come from. Climate advocates claim human activity is causing climate change. Other scientists tell us that methane is a far worst gas in the atmosphere, but it is occurring naturally. Methane is found in shale gas.

Shale gas is a form of natural gas (mostly methane), found underground in shale rock. and it is generally distributed over a much larger area. Typically, shale gas consists of 70 to 90 per cent methane (CH_4), the main hydrocarbon target for oil exploration companies. But there is much more in the universe than what is found on Earth. Look no further than Mars.

CH_4 is found with oceans on Titan, a moon of Saturn. So what does that tell us??? Titan, the largest moon of Saturn, is currently a topic of scientific assessment and research. Titan is far colder than Earth, but of all the places in the Solar System, Titan is the only place besides Earth known to have liquids in the form of rivers, lakes, and seas of methane on its surface. Its thick atmosphere is chemically active and rich in carbon compounds. And the temperature is -270^0 F, so it is doubtful that trees and animals lead to this creation.

A review of other scientific literature leads to the conclusion that increases of CO_2 during the twentieth century have produced no deleterious effects upon global climate or temperature.

Increased CO_2 has, however, markedly increased the growth rates of plants as inferred from numerous laboratory and field experiments. There is no clear evidence of the global effects of CO_2 on climate.

Meaningful assessments of the environmental impacts of anthropogenic CO_2 are not yet possible because model estimates of global in climate on centennial time-scales remain high!

Scientists have been studying our solar system. They have know found that Saturn has more oil and natural gas than Earth and, get this, no evidence of were dinosaurs, plankton and forests on Saturn.

Once again, what do Saturn's moons tell us? Titan has oceans of methane. and why isn't this commonly known and what are the implications of it and what does it tell us about our modern climate change policy.

Willie Soon, an Astro physicist for over 30 years, has been thinking about this for a long time, and has authored and published a number of technical articles about this subject.

So why is eliminating CO_2 better for the planet??? Its not, that is probably why the environmentalists couldn't explain methane and had to **change CO_2 to "green house gasses."**

Of particular interest are comments on how the dark ages started,,, when a clerical elite suppressed common discussion and required blind adherence to a narrowly constrained dogma. Does this sound like the media/government now?

So, maybe "green house gasses." are an integral part of our environment, and are not caused by human activities.

Now is the time to take a look at the polluters and the activities that cause these Green House Gases. China is clearly the largest climate polluter country, making up more than 30% of global emissions. The top 10 global climate polluters are headed by China and India, and the U S is also in the top 10.

Most solar panels are made of silicon, which is the main component in natural beach sand. Silicon is abundantly available, making it the second most available element on Earth. However, converting sand into high grade silicon comes at a high cost and is an energy intensive process. We never hear of the amount of energy needed to produce a solar panel. The process to produce metallurgical grade requires the same amount of power equivalent to using your home oven for 7 hours.

The total picture– most panels come from China, heat at 1/3 the temperature of the sun, from electricity from coal fired power plants, how much pollution???

Currently, 15% of the world's silicon goes into solar panels.

.

Pollution from the Space Race: But there's a lot more to rocket pollution than CO_2. Rockets can produce soot (black carbon), nitrogen oxides, alumina particles, chlorine, hydrochloric acid and water vapors. Rocket launches used to be rare enough that pollution wasn't much of a concern.

Today, according to a study by the National Oceanic and Atmospheric Administration (NOAA), **global rocket launches (of which there were 180 last year**, the study notes) inject about 1,000 tons of soot into the upper atmosphere per year. This will only get worse, NOAA warns, as the space race continues to expand.

Many believe that if rocket launches continue to occur at the current rate that they are – it could provoke the temperature of the stratosphere to rise almost three degrees Fahrenheit while still thinning the ozone layer.

The next big headline will probably be about the stratosphere.

Does this mean that "Tesla and Space X Broke the Sky!"

https://greenly.earth › ecology-news
Mar 10, 2023

https://newsinteractives.cbc.ca

Causes and Solutions Continued:

We all have been to the beach, and standard equipment includes SPF30+++. We need some type of SunProtectionFactor for the earth. Maybe CO_2 will work —- OH,, But that is a greenhouse gas and we need to delete it. Sooo, what about a **beach umbrella???** This is certainly a tried and tested solution. More people need to get on board with this idea! But, there is not enough money for them, I guess.

Could a giant space parasol help solve global warming?

Earth is at its hottest point on record, and experts say humans are not doing enough to stop its overheating. Later, we will take a look at the countries causing the most pollution.

But for now, a small but growing number of scientists are proposing a potential solution that could have leaped from the pages of science fiction: The equivalent of a giant beach umbrella, floating in outer space.

The idea is to create a huge sunshade between the Earth and the sun to block a small but crucial amount of solar radiation — just enough, the scientists say, to keep Earth within manageable climate boundaries.

Does this solution suggest that
Climate Change is a caused by the SUN???
And it may cost less than $7 Trillion

The New York Times <nytdirect@nytimes.com> 2/3/2024

Is this another undocumented claim of the news media and climate activists?

Eddie Current (remember, he's been around for a looong time) reminds us that the source if this photo is not verified., and the timing might, in itself, account for the similarity. Timing like high tide and low tide for example,,,

Global Warming

The Arctic Ocean is warming up, icebergs are growing scarcer and in some places the seals are finding the water too hot, according to a report to the Commerce Department yesterday from the Consulate at Bergen, Norway.

Reports from fishermen, seal hunters and explorers all point to a radical change in climate conditions and hitherto unheard of temperatures in the Arctic zone.

Exploration expeditions report that scarcely any ice has been met as far north as 81 degrees 29 minutes.

Soundings to a depth of 3,100 meters showed the gulf stream still very warm

Great masses of ice have been replaced by moraines of earth and stones, the report continued, while at many points well known glaciers have entirely disappeared

Very few seals and no white fish are found in the eastern Arctic,_while vast shoals of herring and smelts which have never before ventured so far north, are being encountered in the old seal fishing grounds.

Within a few years it is predicted that due to the ice melt the sea will rise and make most coast cities uninhabitable.

I must apologize.

I neglected to mention that this report was from November 2, 1922 as reported by the AP and published in The Washington Post over 100 years ago.

This must have been caused by Model T Ford emissions or possibly from horse farts and what the buffalo did when they once roamed the Great Plains???

Global Warnings are like pizzas. They come in all shapes and forms: some are small or large circles or squares, thin crust, thick crust, stuffed crust, extra toppings.

Causes and Solutions, continued:

BlockChain Background Information:

A blockchain is a public digital ledger which can be used to verify the ownership of digital assets such as cryptocurrencies.

However, many blockchains, and in particular some cryptocurrencies, use a validation mechanism known as 'proof-of-work'. This mechanism requires vast computing power and results in substantial electricity consumption. For Bitcoin alone, more than 100 TWh (TerraWattHours) per year is now consumed, which is equivalent to the annual electricity consumption of the Netherlands.

As of March 2024, it is difficult to provide an exact number of Bitcoin companies as the industry is constantly evolving and new companies are emerging. However, it is estimated that there are 13,200+ companies working within the blockchain ecosystem, ranging from exchanges and wallet providers to mining operations and blockchain development firms.

Did you know that one of the world's first coins had a bee symbol?

There goes my mind!

Let's explores the idea of who's mining bitcoins?

While it's understandable that you might be curious about the impact that aliens mining bitcoins could have on the market, it's important to remember that this is a highly speculative scenario that is not supported by any evidence. Therefore, it would be fruitless to engage in speculation about how such an event might affect the market.

Do you think there's a possibility that Martians are secretly mining bitcoins, given its mysterious value and the vast amount of energy consumption involved?

Digital Mining Energy Costs

The real world cost of digital mining of bitcoin can be quite significant. A major cost is the amount of energy that is required to mine bitcoin. The process involves solving complex mathematical equations, which requires a lot of computational power.

This computational power is provided by specialized computers, which consume a lot of energy.
The energy consumption associated with bitcoin mining is huge. Terawatt-hours per year range are 80+. This equals the energy consumption of an entire
countries like Austria or Chile. This has led to concerns about the environmental impact of bitcoin mining, as it contributes to greenhouse gas emissions and the
depletion of natural resources.

As demand for bitcoin increases, the price of mining equipment also tends to rise, making it more expensive for miners to acquire the necessary tools and resources.

Do you think there's a possibility that aliens are secretly mining bitcoins, given the vast amount of energy consumption involved.

If bitcoin miners were to switch to pedal power instead of using electricity, how many hours of cycling do you think it would take to mine one bitcoin?

Well, that's an interesting question! While it's difficult to give a precise answer, we can estimate the amount of energy required to mine one bitcoin and then determine how many hours of cycling it would take to generate that same amount of energy.

According to recent estimates, it takes approximately 10,000 to 15,000 kilowatt-hours of electricity to mine one bitcoin. Currently, around 19 million bitcoins have been mined and are in circulation.

Assuming that a person can produce around 100 watts of power while cycling, it would take roughly 100,000 to 150,000 hours of continuous cycling to generate the same amount of energy needed to mine one bitcoin. To put that in perspective, that's the equivalent of cycling non-stop for over 11 years!
This is also the equivalent to the amount of energy used by an average American household in a year!

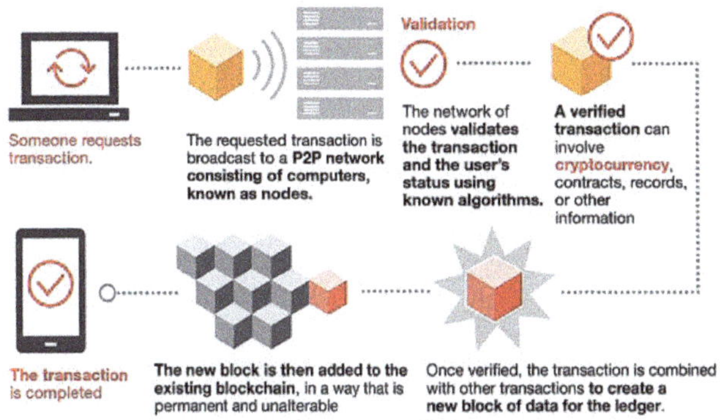

A SIMPLE DIAGRAM OF THE PROCESS

Do you think ***Cycling for Energy Production*** *will become an Olympic Sport???*

Solar panels and wind farms can generate electricity without releasing any greenhouse gas emissions. Nuclear power plants can too, although today's plants generate long-lasting radioactive waste, which has no permanent storage repository. Please keep in mind, with current alternatives, we do not have enough history to know the recycling costs of lithium batteries and wind turbine blades.

Most nuclear reactors can operate for very long periods of time – over 60 years in many cases.

Here is a nuclear power plant in Ohio that was shut down on 2020.

Although they require a lot of maintenance, some nuclear power stations are now certified for 80+ years of operation, far longer than a gas or coal generator. A fact that is not talked about in the
media...

The **problem** is that we are not building any new nuclear plants, *but we can spend $7 trillion on climate change, caused by the power generation plants... Go Figure!*

How is electricity generated using nuclear? - National Grid ESO

Causes and Solutions, continued:

Energy is the centerpiece of most discussions about climate change,,, sooo, Eddie Current is supervising the creation of this document. How is he doing so far?

Recent data shows that electricity usage in the US has increased 28.6%. Another challenge in gaining access all of the renewable energy, the United States needs to dramatically expand the electric grid between places with abundant wind and sunshine and places where people live and work.

More than 20% of the renewable energy installed is **not** connected to the grid. Recent reports show that the U.S. electric grid isn't running 100% of renewable energy yet.

Your guide to the World Economic Forum's annual meeting

Why is Climate Change the second chapter in a detailed study about electricity? Eddie Current has decided to jump into the fight with his vision of all that is happening.

What is Davos and why is it important?

The WEF, (World Economic Forum), has held a meeting every year since it was founded in 1971. But why is Davos, as it is commonly known, so significant?

Most of the year, Davos is unremarkable other than being a popular ski resort high in the Swiss Alps. But for one week in January, it becomes the focus of the world's attention as the global elites all converge on the small alpine town for the annual meeting of the World Economic Forum (WEF). Why? To discuss the future direction of life on our planet and the pressing issues of the day.

By David Walsh Published on 15/01/2024

10 years ago, the production of electricity in the US was 82% based on fossil fuels,,, today, 10 years later, 81% of our electric power is produced by fossil fuels -- at an investment of $7,000,000,000, that's $7 trillion dollars on renewable sources!!!

Reported at the Davos summit Jan 2023

"Take advantage, save thousands of dollars on climate-friendly investments through tax credits and rebates."

Reported March 2024:

Global investment in the energy transition hit $1.8 trillion in 2023, up 17% on the previous year and a new record. The total renewable sources investment now exceeds $8.8 trillion!

Energy Transition Investment Trends 2024 | BloombergNEF (bnef.com)

An early **Climate Investigator** and would-be aircraft designer/inventor took the initiative seeking fame and glory.

Do you think he was trying to measure Methane in the stratosphere???

Lawn chair technology's early success stories start with Larry Walters. Forty-two helium-filled weather balloons lifted him in this aluminum lawn chair from San Pedro, California, on July 2, 1982.

Larry Walters, a.k.a. Lawnchair Larry, rode a lawn chair attached to 42 helium-filled weather balloons in 1982, soaring up to about 4.9 km.

Lawn chair balloon pilot. **Lawrence Richard Walters**, nicknamed "Lawn Chair Larry" or the "Lawn Chair Pilot", (April 19, 1949 – October 6, 1993) was an American truck driver. On 2 July 1982 he attempted to reach outer space in a homemade aircraft.

Facing freezing temperatures and lower oxygen levels, Walters popped balloons with a BB gun to establish a controlled descent. The Federal Aviation Administration charged him with violating controlled airspace, flying without a balloon license, and operating a non-airworthy craft.

Eddie Current's version would most likely state 'A non-Worthy Government foolishly charged a highly creative Citizen with an over abundance of Initiative!'

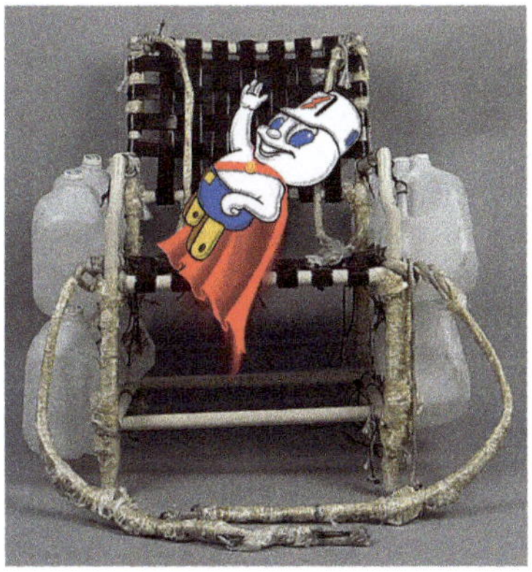

National Air and Space Museum

https://airandspace.si.edu

Chair, Lawn, Larry Walters

Methane produced by "cow burps" is one of the biggest sources of greenhouse gas emissions from agriculture. A single cow produces between 154 to 264 pounds of methane gas per year. Studies have shown that grass-fed cattle produce 20% more methane in their lifetime than grain-fed cattle. Not counting for the emissions of any other livestock, 1.5 billion cattle, raised specifically for meat production worldwide, emit at least 231 billion pounds of methane into the atmosphere each year.

And what about the English & Irish with all their sheep??? https://www.epa.gov › snep › agric...

Disaster Capitalism: A Classic Deep Dive,,,

The exploitation of natural or manmade disasters (such as catastrophic weather events such as, war, epidemics, etc.) in service of capitalists interests: a practice of using unstable social, political, and economic situations to impose or benefit from destruction, the privatization of public interests etc.
Slogan: Burn, Flood, or Shake it to the ground!

Let's zoom in on Chile:

Wildfires cause huge loss of life in Chile in South America

Devastating wildfires have caused dozens of casualties. The disaster occurred as Chile and other parts of South America are gripped by intense heatwaves exacerbated by climate change.

Then add:
Chilean President Gabriel Boric announced preliminary plans for a national lithium strategy, one that would increase government participation in the lucrative sector. Chile is currently the world's second-largest producer of lithium.

Then add:
Currently, only two companies mine lithium in Chile: Chilean firm SQM, which is nearly a quarter-owned by a Chinese company, and U.S. firm Albemarle. They are both private and exclusively operate the country's more than 40 salt flats that are home to lithium reserves.

(*Detailed reference data, like this, can be found throughout)*
https://m.youtube.com/watch?v=pr0LkPMZ-qc
https://youtu.be/QsiGiziM7Vw?si=7ubshkj3YmQHXpRo

(foot notes from Eddie Current)
As Abe Lincoln once said, 'If it is on the internet,,, then it must be true.'

2024, Timely notes & Last minute updates:

#1. When talking about the growth of renewable energy, I often say it's not notable when wind, solar or other technologies reach a record high, because they should be doing that every year. Then 2023 came along. For the first time in more than a decade, the United States had a **decrease in utility-scale electricity generation** from renewable sources, according to the Energy Information Administration.

#2. **Puget Sound Energy** has filed with the Public Utilities Commission for a **20% rate** increase to cover $530 million the **costs of wind turbines**.

#3. **NY attorney general sues** world's largest **beef producer** over methane emissions, climate commitments. NY AG says company's net zero goals aren't feasible considering its high carbon footprint

#4. THE WORLD'S LONGEST ONSHORE **WIND TURBINE BLADE** CREATES A DIAMETER AS LONG AS **THREE FOOTBALL FIELDS** -AND IT'S SET TO DEBUT SOON (That is more than 900ft) The blade, which will be attached to a 15-megawatt wind turbine, is super light and super strong.

#5. Global investment in the energy transition hit **$1.8 trillion in 2023**, up 17% on the previous year and a new record. The total renewable sources investment now exceeds $8.8 trillion!
Energy Transition Investment Trends 2024 | BloombergNEF (bnef.com)

#6. In Georgia, **demand for industrial power** is surging to record highs, with the projection of new electricity use for the next decade now **17 times** what it was only recently. Internet, March 2024

#7. Northern Virginia needs the equivalent of **several large nuclear power plants** to serve all the new data centers planned and under construction. March 2024

#8. Only a measly **1% of US retailers** offer EV charging stations even though they come with significant benefits and government incentives, according to Consumer Reports.

#9. The **Chevrolet Bolt** was the little electric car that could. Never the fastest or the fanciest EV, the Bolt and its sticker price of roughly $30,000 made it cheaper than lots of gas cars, all while delivering a respectable 259 miles of range. With its legion of fans, the Bolt outsold every non-Tesla EV in America last year. But if you want to buy a new one, you'd better hurry: Chevy's parent company, General Motors, stopped making the Bolt at the end of 2023.
Electric Cars Are Still Not Good Enough - The Atlantic

#10. The CEO of **Rivian Automotive** announced Thursday that the electric truck maker is pausing construction of its $5 billion manufacturing plant in Georgia to speed production and save money.
Tesla-rival Rivian is pausing construction on its $5bn electric truck plant in Georgia to speed production and save cash | Fortune

#11. **Our Electric Grids are Blocking a Climate Revolution**
Time.com/6900079/electric-grids-climate-revolution/

#12. Should excessive electricity bills be reported as **a way to identify illegal marijuana** grows in Maine,,, The idea to report high electricity usage comes after frustration from Maine law enforcement who said they can't keep up with the estimated 270 illegal grow operations that have popped up across the state...

POWER COMPANIES

If you lose all power at
your ice-making plant,
do you go
into liquidation?

The following chapter is more than scrapes and bits of information, but, it is enough information for you to change a light bulb and pay your power bill.

Energy & Environment,,,
Electric Utilities in the U.S.

The day starts with breakfast, usually including toast. Where does Eddie Current go when he leaves the toaster? Lets start this chapter with your power company.

An electric utility is most often a company that operates facilities to generate, transmit, and/or distribute electricity to private, public and industrial consumers. Electric utilities can be involved in one or more of these activities. Some utilities only function in several categories. TVA, for example, only generates and distributes electricity. Others, like electricity marketers, who only buy and sell electricity, can also be considered utilities.

In the United States, there are around **3,000** electric utility companies providing power to more than **140 million customers.**

Too much detail???

More details are coming,,,

Published by Statista Research Department, Dec 18, 2023

Found on the internet: The world is at a turning point where it must switch from fossil fuels to renewable energy sources in order to tackle climate change. Because they are clean, plentiful, and sustainable, renewable energy sources like wind and solar energy are gaining popularity. Yet, one of the problems with renewable energy is the intermittent nature of these sources and thir energy storage.

"We need to eliminate global emissions of greenhouse gases by 2050," philanthropist and technologist Bill Gates wrote in his 2023 annual letter...

Solar energy systems can only generate electricity during the day, while wind turbines can only do so when there is enough wind. Because of this, it is challenging to rely on these sources for a regular and reliable energy supply.

Thankfully, technology has improved and a variety of renewable storage options are now readily available. We will examine the various forms of renewable energy storage on the following pages.

https://www.iea.org/commentaries/tripling-renewabke-power-capacity-by-2030-is vital-to keep-the -150c-goal-within-reach

WOW, 27 years and $7 trillion,,, feel the sense of urgency yet???

Lets Take a Look at Your Local Power Company

I hope that the following pages will help shed some light on the complex subject of your power bill. But, if you want to skip these pages, please feel free.

As we look at the charges for the work that Eddie Current does for us across the country, we can see a great variety in charges and rational for providing this service. Some localities have simple, easy-to-understand billing formats, while others desire to complicate the entire process—probably so that they can get away with charging more.

We have selected several states that are typical for their practices and geography. Using the universal measure of quantity, the Kilowatt Hour (KWH), we have demonstrated a range of charges of $0.136/ KWH to $0.359/KWH (13.6 ¢/KWH to 35.9¢/KWH).

California has unusually high electricity rates because of state mandates for sustainable energy, mostly hydro-electric from Grand Coulee Dam and the Pacific Northwest. There has been serious under-investment in conventional nuclear and fossil fuel electricity in favor of solar and wind subsidies. It is the most expensive electricity in the country.

Lets remind ourselves that Eddie Current is the same guy in every state. The work he does is the same powering lighting, computers, cooking, and TVs. He comes from the same sources of fossil fuels, solar, and wind, but your charges vary greatly.

I hope that the following pages help shed some light on this complex subject.

As we look at the charges for the work that Eddie Current does for us across the country, we can see a great variety in charges and rational for providing this service. Some localities charge have simple, easy-to-understand billing formats, while others desire to complicate the entire process—probably so that they can get away with charging more.

Here are a few of the highlights of the complex details Included are:

~Fuel and Non-fuel charges **(?),**

~Delivery charges and Transition charges ***(whatever those are),***

~Merchant function charges ***(why),***

~Peak and Off-Peak charges ***(I think I can understand this one),***

~Court resolution surcharge (***?***),

~Meter reading ***(that's the basic cost of doing business!),***

~Regulatory assessment (***?***),

~Franchise fee ***(give me a break—this is not a McDonalds)***

Please remember, the universal measure of quantity is the Kilowatt Hour (KWH), (which we have never seen, let alone even measured and we have demonstrated a range of 13.6 ¢ to 35.9 ¢ for the same KWH.

#1. **This New York State Electric & Gas** power bill from upstate New York is quite confusing,,, I had to read it several times... I believe that the rate in this area averages out to $0.178/KWH... (17.8 ¢/KWH)

And it seems that a lot of government agencies get a piece of the action.,., including, but not limited to, 'transition charges,' 'merchant function,' and 'school dist. sales tax,,,' *whatever these charges mean...*

***Electricity Service -** Residential* *Service from: 12/22/23 - 01/19/24*
***Electricity Rate -** 12001 NYSEG Supply Service* *Po*

Meter Number	Current Meter Read Date	Reading	Previous Meter Read Date	Reading	Reading Difference	Billed Usage	Billing Period
00	01/19/24	57450 A	12/22/23	56669 E	781	781 kwh	29 days

Type of read: A - Actual, E - Estimate, C - Customer, R - Remote and N - No read

025584 2/2

Electricity Delivery Charges

Basic service charge					19.00
Delivery charge	781	kwh	@	0.06415	50.10
Transition charge	781	kwh	@	0.00445478	3.48
Revenue decoupling mech	781	kwh	@	-0.000816	-0.64
SBC charge - Dec	255	kwh	@	0.004933	1.26
SBC charge - Jan	526	kwh	@	0.004616	2.43
Subtotal Electricity Delivery					$75.63

Electricity Supply Charges

Supply charge	781	kwh	@	0.07[illegible]9276	56.18
Merchant function charge - Dec	255	kwh	@	0.002[illegible]85	0.74
Merchant function charge - Jan	526	kwh	@	0.00287[illegible]	1.51
Subtotal Electricity Supply					$58.43

Electricity Taxes and Surcharges

Taxes on delivery charges	@	2.0408%	1.54
School district sales tax	@	2.5000%	3.39
Subtotal Electricity Taxes and Surcharges			$4.93

Total Electricity Cost	**$138.99**
Total Energy Charges	**$138.99**

Usage Chart Information

Historical Electricity Consumption

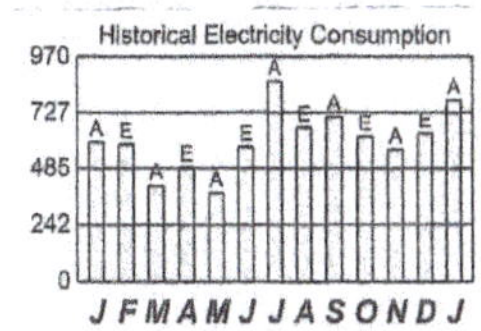

Electricity Daily Average Comparisons

Billing Period	Average Daily Use	Average Daily Temp
Jan-24	27 kwh	31° F
Jan-23	18 kwh	31° F

Miscellaneous Charges

01/24/24	Payment & billing svcs charge sales tax	0.02
01/24/24	Payment & billing svcs charge	0.89
01/24/24	Payment & billing svcs GRT	0.02
Total Miscellaneous Charges		**$0.93**

FPL

Customer Name: **Account Number:**

FPL.com Page 2

BILL DETAILS

Amount of your last bill		196.24
Payment received - Thank you		-196.24
Balance before new charges		$0.00
New Charges		
Rate: RS-1 RESIDENTIAL SERVICE		
Base charge:	$[illegible].48	
Non-fuel: *(First 1000 kWh at $0.087920) (Over 1000 kWh at $0.097840)*	$88.69	
Fuel: *(First 1000 kWh at $0.034620) (Over 1000 kWh at $0.044620)*	$34.98	
Electric service amount	133.15	
Gross receipts tax (State tax)	3.42	
Taxes and charges	3.42	
Regulatory fee (State fee)	0.10	
Actual electric charges	136.67	
Budget billing charges		
Total amount you owe		

FPL automatic bill pay - DO NOT PAY

EDI File Transmitted Separately

METER SUMMARY

Meter reading - Meter ACD1506. Next meter reading Feb 21, 2024.

Usage Type	Current	-	Previous	=	Usage
kWh used	76970		75962		1008

ENERGY USAGE COMPARISON

	This Month	Last Month	Last Year
Service to	Jan 22, 2024	Dec 20, 2023	Jan 21, 2023
kWh Used	1008	1110	1095
Service days	33	30	32
kWh/day	31	37	34
Amount	$136.67	$152.12	$138.33

FPL BUDGET BILLING

Deferred Balance $54.52

KEEP IN MIND

- Taxes, fees, and charges on your bill are determined and required by your local and state government to be used at their discretion.
- The fuel charge represents the cost of fuel used to generate electricity. It is a direct pass-through to customers. FPL does not profit from fuel, although higher costs do result in higher state and local taxes and fees.

#2. Florida Power & Light...

Here's a relatively easy electric bill to understand. Only 3 line items, Base, Non-fuel, and Fuel.

I believe that the rate in this area averages out to $0.136/KWH...(13.6 ¢/KWH)

ENERGY STATEMENT
www.pge.com/MyEnergy

Account No:
Statement Date: 01/10/2024
Due Date: 01/31/2024

Service For:

Questions about your bill?
Mon-Fri 7 a.m.-7 p.m.
Saturday 8 a.m.-5 p.m.
Phone: 1-800-743-5000
www.pge.com/MyEnergy

Ways To Pay
www.pge.com/waystopay

Your Account Summary

Amount Due on Previous Statement	$292.12
Payment(s) Received Since Last Statement	-292.12
Previous Unpaid Balance	$0.00
Current PG&E Electric Delivery Charges	$160.50
Central Coast Community Energy Electric Generation Charges	64.93
Total Amount Due by 01/31/2024	**$225.43**

#3. Pacific Gas & Electric

I have been doing research on this phenomenon called electricity now for more that 45 years and I find that Eddie Current is the same guy in California as in New York or Florida… But look at his wages, thru Pacific Gas & Electric— $0.359/KWH. (35.9¢/KWH)

But it is much more complicated. His actual charges are …10.3¢/KWH and his fringe benefits total. 25.5¢/KWH, but vary depending on the 'time-of-day',

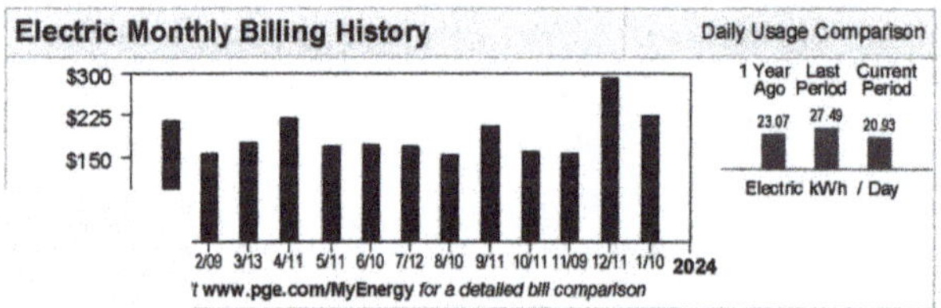

Total energy,,, 4 different charge rates. Why?

ENERGY STATEMENT
www.pge.com/MyEnergy

Details of Central Coast Community Energy Electric Generation Charges
12/11/2023 - 01/09/2024 (30 billing days)
Service For:
Service Agreement ID:
Rate Schedule: MBRETCH1 3Cchoice ETOUB

12/11/2023 – 01/09/2024

Generation - Peak - Winter	64.553100	kWh	@ $0.15700	$10.13
Generation - Peak - Winter	40.659200	kWh	@ $0.18300	7.44
Generation - Off Peak - Winter	326.240900	kWh	@ $0.08500	27.73
Generation - Off Peak - Winter	196.341900	kWh	@ $0.09900	19.44
Energy Commission Tax				0.19
Total Central Coast Community Energy Electric Generation Charges				**$64.93**

Account No:
Statement Date: 01/10/2024
Due Date: 01/31/2024

Your Electric Charges Breakdown (from page 2)

Transmission	$32.35
Distribution	102.74
Electric Public Purpose Programs	16.53
Nuclear Decommissioning	-0.09
Wildfire Fund Charge	3.40
Recovery Bond Charge	3.31
Recovery Bond Credit	-3.31
Wildfire Hardening Charge	1.59
Competition Transition Charges (CTC)	0.36
Energy Cost Recovery Amount	-0.29
PCIA	3.19
Taxes and Other	0.72
Total Electric Charges	**$160.50**

ENERGY STATEMENT
www.pge.com/MyEnergy

Account No:
Statement Date: 01/10/2024
Due Date: 01/31/2024

Details of PG&E Electric Delivery Charges
12/11/2023 - 01/09/2024 (30 billing days)
Service For:
Service Agreement ID:
Rate Schedule: ETOUB B Residential Time-of-Use Service

Service Information

Meter #	
Total Usage	627.795100 kWh
Heat Source	B - Not Electric
Serial	Q
Rotating Outage Block	50

12/11/2023 – 12/31/2023

Energy Charges				
Peak	64.553100	kWh	@ $0.39763	$25.67
Off Peak	326.240900	kWh	@ $0.35883	117.07
Generation Credit				-52.58
Power Charge Indifference Adjustment				1.03
Franchise Fee Surcharge				0.43

01/01/2024 – 01/09/2024

Energy Charges				
Peak	40.659200	kWh	@ $0.46592	$18.94
Off Peak	196.341900	kWh	@ $0.42712	83.86
Generation Credit				-36.37
Power Charge Indifference Adjustment				2.16
Franchise Fee Surcharge				0.29
Total PG&E Electric Delivery Charges				**$160.50**

2018 Vintaged Power Charge Indifference Adjustment

11 different charges. Why???

4. Arizona Public Service,,,

I believe that the rate in this area averages out to $0.172/KWH... (17.2¢/KWH),,, and here's how the bill breaks down: **18 line** items for Eddie Current, **5 line** items for governmental getting their fingers into this activity,
4 line items for time-of-day usage.

It is a conspiracy to undermine and attack our level of understanding.

I hope that these pages help shed some light on this complex subject. As we look at the charges for the work that Eddie Current does for us across the country, we can see a great variety in charges and rational for providing this service. Some localities charge have simple, easy-to-understand billing formats, while others desire to complicate the entire process—probably so that they can get away with charging more. The universal measure of quantity is the Kilowatt Hour (KWH), and we have demonstrated a range of $0.136/KWH to $0.359/KWH.

And increases are in the pipeline, so to speak!

Your electricity bill **Morel** **Your account number**
January 9, 2024

Service plan: Time-of-Use 4pm-7pm Weekdays

Meter number:
Meter reading cycle: 06

Charges for electricity services

Cost of electricity you used

Customer account charge	$2.18
Super Off-peak delivery service charge	$0.91
On-peak delivery service charge	$1.23
Off-peak delivery service charge	$8.84
Environmental benefits surcharge	$3.57
Federal environmental improvement surcharge	$0.06
System benefits charge	$1.33
Power supply adjustment*	$8.07
Metering*	$6.02
Meter reading*	$2.18
Billing*	$2.43
Generation of electricity super off-peak*	$0.60
Generation of electricity on-peak*	$9.75
Generation of electricity off-peak*	$18.91
Federal transmission and ancillary services*	$4.64
Federal transmission cost adjustment*	$0.16
Court resolution surcharge	$0.74
LFCR adjustor	$1.11
Cost of electricity you used	$72.73

Taxes and fees

Regulatory assessment	$0.18
State sales tax	$4.16
County sales tax	$0.52
City sales tax	$0.74
Franchise fee	$1.46
Cost of electricity with taxes and fees	$79.79
Total charges for electricity services	**$79.79**

** These services are currently provided by APS but may be provided by a competitive supplier.*

Amount of electricity you used

Meter reading on Jan 9	20426
Meter reading on Dec 8	20003
Total electricity you used, in kWh	**423**
On-peak meter reading on Jan 9	3586
On-peak meter reading on Dec 8	3545
On-peak electricity you used, in kWh (4pm - 7pm Monday - Friday)	**41**
Off-peak electricity you used, in kWh (All other hours and certain holidays)	**296**
Super off-peak meter reading on Jan 9	4039
Super off-peak meter reading on Dec 8	3953
Super off-peak electricity you used, in kWh (10 am to 3 pm Monday - Friday, November - April)	**86**

Average daily electricity use per month

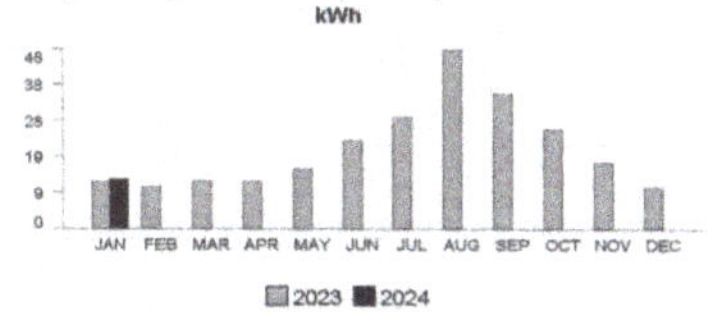

Comparing your monthly use

	This month	Last month	This month last year
Billing days	32	30	33
Average outdoor temperature	54°	63°	51°
Your total use in kWh	423	334	412
Your average daily cost	$2.49	$2.28	$2.22

Your Monthly Plan Comparison View more details or switch plans by calling (800) 253-9405 or visiting **aps.com/compare.**

Your Current Plan	Save The Most With	This Month	12 Month Savings
Time-of-Use 4pm-7pm Weekdays	**Fixed Energy Charge Plan - Tier 2**	**$6.29 more**	**$61.22**

What we need is more agencies involved in this power industry.

The electrical power industry has evolved significantly over the years, with the emergence of various market participants and channels of distribution.

COMPLEXITY? YEP!!!

Eddie Current first was recruited by the TVA (Tennessee Valley Authority) as a Marketing Manager, which started out as a unified industry where the electrons were produced and then distributed to Municipal Power Companies and REAs (Rural Electric Authorities). The electrons went from the producer, were transmitted directly to the distributor, who then delivered directly to our homes and businesses... A simple 3 step supply chain.

Then Government got its sticky fingers into the pot. The next step in this process was adding the step referred to as 'Transmission.' This was necessary because the 'plant' was located so far from where we lived and worked.

So, initially, there were three elements in the electron process— Generation,
Transmission,
Distribution.

Federal and local governments quickly saw an opportunity to get a piece of the action and then created Public Utility Commissions (PUC).

Firstly, it is important to note that the electrical power industry has undergone significant changes since the early 1990's. The deregulation has led to the emergence of various market participants.

Get the picture now???
Maybe it is like wrestling a greased pig!

Here's what the industry looks like today...

1. **Generators:** Where electrons are produced - coal-fired power plants, gas-fired power plants, nuclear power plants, hydro plants, wind farms, and solar farms.

2. **Transmission System Operators (TSOs):** Organizations that operate the high-voltage transmission network giving Eddie Current a path to your neighborhood. They can also be referred to as independent system operators **(ISOs)** and regional transmission organizations **(RTOs)** who manage grid operations

3. **Distribution System Operators (DSOs):** Organizations that deliver electrons to your door. Also known as Grid Operators. There are actually 10 covering the USA

4. **Public Utility Commissions (PUC):** Government agencies that give all players a chance to make HUGE profits.

5. **Retailers or Suppliers:** Organizations that buy electricity from the wholesale market and sell it to those who need it.

6. **Aggregators:** These are entities that bundle power supply from multiple sources and sell it to end-consumers.

7. **Brokers:** Intermediaries between buyers and sellers.

8. **Market Operators:** Entities that provide a trading platform.

We find upon closer examination that each of these segments has more than a few sub-categories!

What is the Grid?

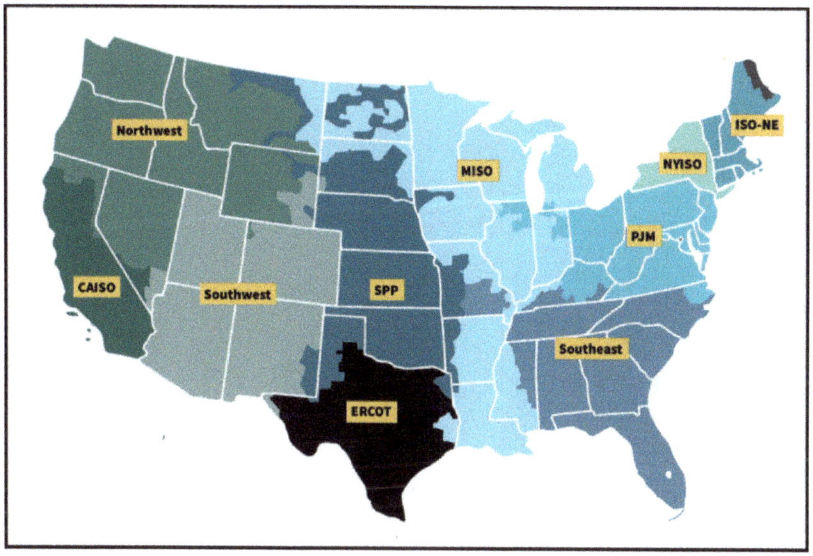

The Grid links many states to insure a more reliable system of supply. Texas is a bit different, and ERCOT was formed. What Is ERCOT???

The North American electric power system is configured with 3 grids, The Eastern Interconnection, The Western Interconnection. The Texas Interconnection, the Electric Reliability Council of Texas (ERCOT) is independent.

In many parts of the country, independent system operators **(ISOs)** and regional transmission organizations **(RTOs)** manage grid operations. **ISOs** grew out of the Federal Energy Regulatory Commission **(FERC),** issued on April 24, 1996. In the orders, the commission suggested the concept of an **ISO** as one way for existing tight power pools to satisfy the requirement of providing non-discriminatory access to transmission. Subsequently, FERC Order No. 2000, issued on Dec. 20, 1999, allowed the voluntary formation of RTOs to administer the transmission grid on a regional basis throughout North America, including Canada.

While major sections of the country operate under more traditional market structures, two-thirds of the nation's electricity load is served in RTO regions. Among the system operators are California ISO (CAISO), Midcontinent ISO (MISO), ISO-New England (ISO-NE), New York ISO (NYISO), Southwest Power Pool (SPP), PJM Interconnection, and Electric Reliability Council of Texas (ERCOT). Source: FERC (map retrieved Dec. 29, 2023)

*—**Aaron Larson** is POWER's executive editor (@POWERmagazine).*

Number of electricity providers in the United States in 2022, by ownership type

Characteristic	Number of utilities
Investor owned	**166**
Political subdivision	83
Wholesale power marketer	37
Community choice aggregator	26

*What is factor (**E**yes*

*your **EGO** now,,, 10??? **G**lazed **O**ver)*

How many coal-fired power plants are there in the world today?

I think it's pretty cool how Asian people made a language entirely out of tattoos.

Great question. These are the plants world wide that are supposedly causing the most climate change and are driving the race to renewable generation.

The EU has 468 - building 27 more... Total of 495
Turkey has 56 - building 93 more... Total 149
South Africa has 79 - building 24 more... Total 103
India has 589 - building 446 more... Total 1035
The Philippines has 19 - building 60 more... Total 79
South Korea has 58 - building 26 more... Total of 84
Japan has 90 - building 45 more... Total 135
China has 2,363 - building 1,171 more... Total = 3,534

That's 5,615 estimated coal-powered plants in just 8 countries.

USA has 15 - building 0 more...Total = 15

However, in the meantime, China still consumes nearly five times as much coal as India, and nearly six times as much as the United States and is building a huge number of new plants.

China is Currently Building Over Half of The World's New Coal-based Power Plants.

In 2021, China began building 33 gigawatts of coal-based power generation, according to the Helsinki-based Center for Research on Energy and Clean Air (CREA). That is the most new coal-fired power capacity China has undertaken since 2016 and, says CREA, three times more than the rest of the world combined. India, Brazil and Russia are right behind China.

The gears of Climate Change grind slowly, and are driven by electrons.

China Has Built 14 Overseas Coal Plants Since Vowing No New Ones

https://www.bloomberg.com/news/articles/2022-09-22/china-has-built-14-overseas-coal-plants-since-vowing-no-new-ones

https://e360.yale.edu/features/despite-pledges-to-cut-emissions-china-goes-on-a-coal-spree

Chinese Coal Based Power Plants

Industrialization was encouraged by the ready availability of plentiful and abundant cheap energy/electricity here in the US. Here is an example how China is funding its growth in industrializing. Smoking chimneys and a solar panel farm are seen along a highway in a coal producing region in Yulin in northwestern China's Shaanxi province, April 24, 2023.

Europe has its own unique energy problems. After years of misguided energy policies, Europe's electricity has become so expensive that trade unions have started warning of the threat of deindustrialization. Many of its nuclear plants are being shuttered and new sanctions are disrupting the supply of Russian natural gas. Additionally, the European Union has doubled down on renewable energy mandates, further constricting the supply of fossil fuel power.

Chinese need cheap power to make tee shirts and plastic toys!

Europe's electricity prices have settled at triple their pre-pandemic levels, and are projected to remain at this level for some time.

Will this be the unforeseen consequence here in the US???

High electricity prices have Europe facing deindustrialization; don't let it happen here | The Hill OPINION-ENERGY AND ENVIRONMENT—Jan 2024

The trend so far looks very different in countries with **rapidly growing economies** — nowhere more so than in China.

China overtook the United States as the world's single largest power producer in 2010, and now makes nearly a third of the world's electricity. (The country's per person electricity generation is still much lower than America's.)

(Where did all these electrons come from??? Did Eddie Current's family suddenly grow? Where are all these electrons buried in the ground?)

For decades, the country's soaring power demand was fulfilled largely by coal, the most polluting fossil fuel. Coal-fired generation continued to grow, though at a slower pace, even as China *significantly expanded carbon-free power* in recent years, especially wind and solar.

Energy analysts expect carbon-free power to grow enough in the next few years to start pushing out coal-fired electricity here, too.

How quickly coal power and related emissions will decline after that peak is less clear. In a recent agreement with the United States, China said it would speed up with able energy deployment to "accelerate the substitution for coal, oil and gas generation" and agreed to pursue "meaningful" power-sector emissions cuts over the next decade. But whether it will stop approving new coal plants remains to be seen.

Electricity Around the World

By Nadja Popovich Nov. 20, 2023 New York Times

Carbon-free electricity has never been more plentiful. Wind and solar power have taken off over the past two decades, faster than experts ever expected. But it hasn't yet been enough to halt the rise of coal and gas burning generation.

That's because global demand for electricity has grown even faster than clean energy, leaving fossil fuels to fill the gap. The dynamic has pushed up carbon emissions from the power sector at a time when scientists say they need to be falling — and fast — to avoid dangerous levels of global warming.

A whole bunch of folks just aren't listening or maybe they just don't care!!!

There are countries where electrical connections are really scary???

Everybody uses the pole space,,, Electric power, Telephone, Internet, Cable TV, Wired Security companies, and Street lights are often mounted. The wiring is not even color coded!

"I FINALLY MANAGED TO GET RID OF THAT
NASTY ELECTRICAL CHARGE
I'VE BEEN CARRYING...
I'M EX-STATIC!"

Do you think the local infrastructure needs upgrading?

Do you think this was planned, or did it just happen?

An early theory on adding new electricity generation that still might grow on you.

Mother Earth News

If you plant a light bulb in your garden, does it grow into a power plant???

A power generation plant in Florida with customer friendly solar panels in the parking lot just grew out of the landscape...

Eddie Current is a 'big fan' of wind turbines.

'Not In My Backyard'
Home values depreciate about 10% when view of wind turbines is less than 5 miles away. The total loss in values across all US houses with a view of windmills adds up to a drop in the US of $24.5 billion

ELECTRIC VEHICLES

BATTERY
DISCONNECT
CHARGING
PORT

veo

trip BIKES

Trip Bikes is a new company headquartered in Columbus Ohio. Their focus is on Personal Mobility in the form of creation on an all new *e*-bike. Trip software-enabled E-bikes have an innovative software platform. This will allow Trip to take *E*-bikes to the next level and accomplish things like virtual parking and Trip tours.

Rise of micromobility: The global micromobility market is worth about $180 billion today. McKinsey analysis shows that the value could more than double by 2030 to reach about $440 billion.

MicroMobility, driven by the Gen Z segment of our population is the major motivating factor, and of course it has to be powered by Eddie Current.

In the US, there are about 250 E-bike brands currently selling bikes, in the EU, by comparison, there are nearly 500 ebike brands.

The total E-bike market in 2022 was in excess of $48 billion.

Rising demand for E-bike not only for recreational activities but also for daily commute purposes, growing environmental concern for reducing carbon emissions, and increasing supportive programs by the U.S. government encouraging the creation of bike highways which has been started in many cities are the key factors driving the market growth.

How an electric bike is classified:

Class 1: Top speed 20 mph; pedal assist only; allowed on bike paths and MTB trails in most states, with some restrictions
Class 2: Top speed 20 mph; pedal assist and throttle modes; allowed on bike paths and MTB trails in most states, with some restrictions
Class 3: Top speed 28 mph; pedal assist mode (to top speed), throttle mode (limited to 20 mph); allowed on streets and bike paths; off-road use restricted in most states to motorized trails only.

Let me think,,,
At my age,
Should I try this
or not?

***Indecision* is the key to flexibility...**

4 Wheels are much safer, and at my age, safety is a major driver!

What kind of car does an electrician drive?

A "Volts-wagon."

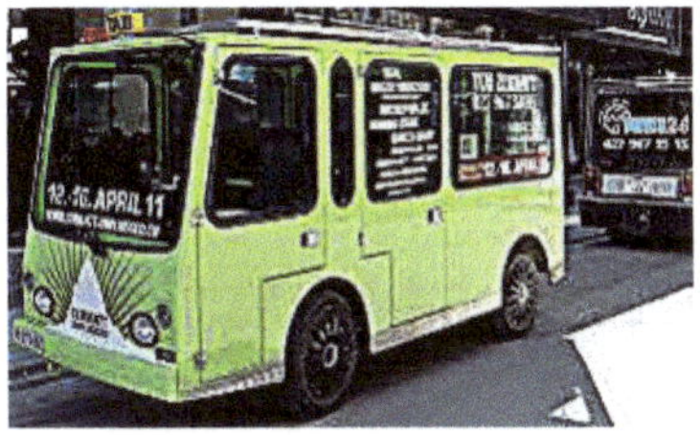

EV Manufacturing Facts:

BYD, a Chinese conglomerate, is the largest EV company based on the nearly 1.9 million EVs it manufactured in 2022, according to EV-Volumes. About half of those were plug-in hybrid EVs (PHEVs), and the other half were battery electric vehicles (BEVs).

Tesla built 1.3 million EVs in 2022, the most by a U.S. company and second most in the world. Tesla built the highest number of battery electric vehicles in 2022.

What is Electricity & How Does It Make The Car go?

Are the kilowatts that drive your new Tesla the same if they come from a coal fired power plant or from a source like a solar panel or wind turbine? **Lets ask Eddie Current:**

"Well Mister Ed, here's my story. When a new car is delivered it has "a full tank," if you know what I mean. My buddies and I are stationed in all four tires just as you see us now. Then you give us some "juice," by stepping on the "go" pedal.

We are now ready to do our job. However, when you make us do a lot of work, we just plain run out of juice, we get too pooped to pop. I mean, how much energy would you have left over if you had to pull a 2 ton car another 300 miles on an empty belly? Even with a bunch of friends... We're no different then you...

Here is how we work. We hop *into* the tires and RUN round and round—*like a hamster on a treadmill,,, (engineers call this a motor).* Thus the car goes forward, if in reverse, then I am recharging myself.

When we get tired, we need to be plugged in to a charging station which is merely giving us the energy we need to continue to do the work. Actually the 'battery' is really a lithium hot-tub, and plugging in to a charging station only heats up the hot tub which energizes us. But luckily for you, instead of needing beef, chicken, chicken thighs, chicken fat, salmon, sheep guts, and fish, we are happy to just suck the heat from the lithium.

Just a thought,,, *When thinking back about the past in transportation imagine when everyone rode horses and only the rich had cars, and now today, everyone owns at least 1 auto and only the rich ride horses... imagine the future world where all cars were EVs, then along came the 'Internal Combustion Engines.'*

Would they sell like hotcakes? Advertising would say: Here is a vehicle that is half the weight, half the price, that will cause less than 1/4 the damage to the roads, that could be refueled in 1/10 the time, with a range up to 4 times and had no limitations due to weather.

Maybe, most importantly, its manufacturing does not demand environmentally damaging non-renewable rare earth elements,,,

The battery pack (Lithium Hot-Tub) on the typical EV

Could this be,,,
What are we hearing sounds like the shouting's of an aluminum siding salesman?

As **electric vehicle demand slows,** workers caught in the middle face an uncertain future

This year was shaping up to be a good one for the workers at the GM Orion Assembly plant in a working class suburb of Detroit. After winning a pay raise following last year's United Auto Workers strike, they were slated to start production later this year on a marquee product for GM: the electric Chevy Silverado pickup truck.

But like thousands of other workers on the front lines of the electric vehicle transition, they have hit some bumps in the road. GM told the nearly 1,000 workers at the Orion plant in December that they were being laid off until late 2025 to make engineering improvements amid cooling demand for electric vehicles. Their last paycheck was their holiday pay the week of Christmas, and many are still waiting to find out if they will be offered a job at another plant.

"It's been a very somber moment here the last month," said an employee at the plant who asked not to be identified because he wasan't authorized by GM to speak to the media. "We thought we finally got a little bit of a break. I think the automotive industry is hurting. In my opinion, they put the carriage in front of the horse."

"Ultimately it all stems from demand, and demand is just not showing up to where all these CEOs thought. So a lot of the initial targets put out by GM or Ford a couple of years ago have maybe proven to be a bit too optimistic and probably too aggressive," said Gabe Daoud, a sustainable energy senior analyst at TD Cowen. "I think everybody was expecting the entire car fleet to change overnight and go electric, but that's obviously just impossible and impractical."

Along with GM, Ford said in January it is cutting back on production of its electric pickup truck the F-150 Lightning amid slowing demand, eliminating 1,400 workers from the production line in Dearborn, Michigan, starting in April.

Someone in the Traffic Department has a great sense of Humor!

https://www.motortrend.com/reviews/2022-lucid-air-grand-touring-performance-yearlong-review-update-1-cold-weather-range-test/

What Is MPGe? The New Fuel Efficiency Rating for Hybrids and EVs

Here is an interesting fact:

Everyone knows what a gallon of gasoline looks like,,, how many times have you filled up a can at the gas station for the lawn mower or boat but, what does a kilowatt hour of electricity look like??? How many "D" batteries does it take to make a kilowatt hour of electric power? This might be a featured topic for a 2nd Edition of "What is Electricity and where does it go when it leaves the pump?"

Let's get back to the current story. Yes, you are correct, Eddie Current is running on a treadmill, just like a hamster. This produces the energy to power an EV.

It is also great training for running inside the tires. This is actually the activity which creates the torque that makes the EV go forward! Then when he turns around, you have put the selector into reverse, and you are putting 'juice' back into the battery.

"It's actually electric. The only bad thing is you can only go 50 feet before you have to charge it again."

An EV takes 40% less labor to produce, think about that— only 60% of the labor to produce a more common gas powered vehicle. And In fact, according to Tesla, their drivetrain only has about 17 moving parts compared to the 200 or so in a typical drivetrain for an internal combustion driven auto.

Sooo, why the high cost of an EV??? Or better yet, why the duplicity of our government in their subsidizing and pushing a 'Green Agenda'.

Everyone knows what a gallon is. We experience it almost daily. When we go to the grocery store and buy a gallon of milk or a gallon of Arizona Tea. And, most commonly, when we fill up our old internal combustion engine cars the meter on the pump tells us how many gallons, down to the tenths, was pumped into the tank.

But now our government wants us to give them complete trust when they say that the energy in one gallon of gasoline is equal to 33.7 kilowatt hours of electricity. This is very close to the magicians slight of hand trick. Are we comparing to regular or premium gas? What octane rating? E85 or diesel?

Who has seen a kilowatt hour of electricity, let alone 33.7 of them in one place? Who do you trust? The savings and annual fuel cost are in the largest type font,,, while cost of a KWH is at the bottom where nobody ever reads that far down and is so small you can't even see it,,, $0.13 cents per kilowatt hour, *reality* only in your wildest dream.

With this seemingly sound criteria, the government created an evaluation terminology called MPG*e,* representing the number of miles the vehicle can go using a quantity of fuel with the same energy content as a gallon of gasoline, so it says on the EPA website.

95% of all electric vehicles are still on the road... The remaining 5% made it all the way home.

Public charging stations are becoming more readily available, many thnx to government sponsored incentives.

Tesla - Toasters - Teslas

Here's the latest news on **EV's in the US,,,**

Adding insult to injury for the EV brands is the news that Hertz is selling 20,000 EVs, or one-third of its total U.S. EV fleet, (because of high maintenance costs and the inability to rent at reasonable prices) in favor of purchasing gas-powered vehicles and incurring $245M of depreciation expense in the process.

one-third of Hertz's total U.S. EV fleet sold and replaced by gas powered vehicles.

January 11, 2024 12:39 PM EST) CFRA analyst Garrett Nelson

What do you call a Rooster staring at a piece of lettuce???

A Chicken-Sees-A-Salad!!!

Crazy starts with a **"C",** and so does **C**alifornia... Several cities in California have now passed laws stating that the police cannot give traffic tickets to Electric Vehicles (EV's.) So, when you are in your Tesla, or Lucid, and you are in an accident and hear the sirens coming up behind you, your best bet is hop into the back seat! You're safe and the auto can't get a ticket!

We're happy you could contribute to our Policemen's Ball,,, Here's your coupon for a free bucket of Eddie Currents on your next fill-up,,,

(Author's Note: It is the reader's option to take these next few pages seriously)

Electric Cars: A plan that doesn't work to fix a problem that doesn't exist.

I just got a full tank of gas for $22.57. Granted, It was for my lawn mower.

Outdoor Temperature, -4^0 F

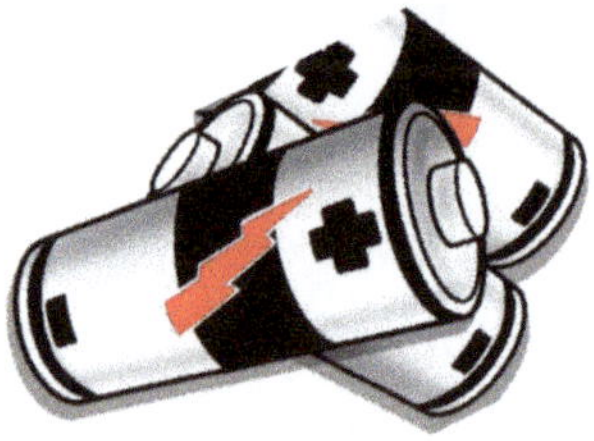

So when these Electric Car's batteries overheat and burst into flames, is it still considered a zero emissions car? Asking for a friend.

My day is complete.

A gas powered generator for an electric car.

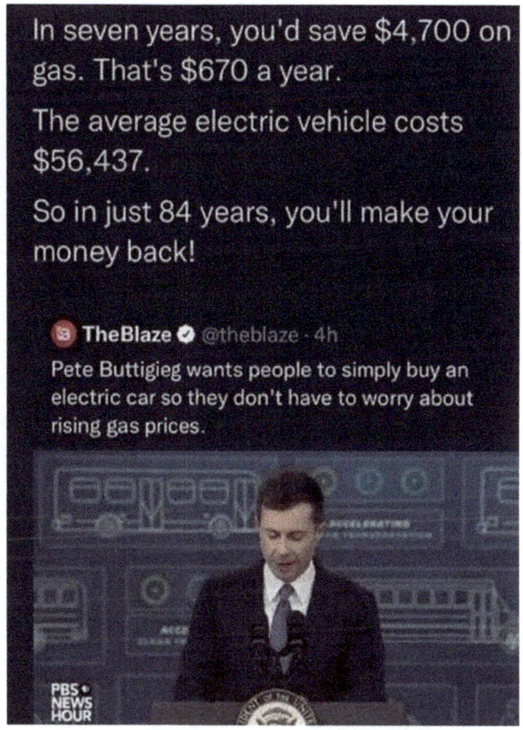

More than 284 million vehicles (cars, motor-cycles, trucks, & buses) are registered in the US.

The number of cars in the world is estimated at more than 1.4 billion.

Europe has about 400 million.

And, reportedly, China gets the trophy as having the most autos (415 million).

How to write "I changed a lightbulb" on your resume:

"Single-handedly managed the successful upgrade and deployment of a new environmental illumination system with zero cost overruns and no safety incidents."

If you get sick riding in an EV, would that be diagnosed as an Auto-Immune disorder???

Here we are,,, into the mid-2020's,,, millenniums after these elements arrived in our universe,,, and they are still here,,, we have no shortage of the **sun, and tornados and lightning!**

And now, we have a whole bunch of people telling us how we should live our lives...

When I asked "Why," when I was a youngster, My grandfather often told me ~~~ *'If you don't understand something, then there is money behind it.'*

From an anonymous Wisconsin State Trooper

I'm not a fan of electric vehicles. Too many variables affecting battery consumption. Definitely not suited for cold climates. The following experience just cements my distaste for EV's, especially Teslas:

I got sent to a motorist assist the other day, at the start of our snowstorm. Tesla on the side of the interstate, dead battery. So, I arrive on scene and the occupants have the right-front door open. They tell me that they can't open any other doors, because the battery is dead. Sure enough. Can't open the doors from inside or outside. The driver also can't get her license out of the glove box where she put it during their trip. Because the glove box opens electronically... and the battery is dead. You actually have to use the computer in the center of the dash to open the glove box.

They said they had 10% battery left, should've been plenty to get from that location to the charging station nearby. Then all of a sudden, the whole car shut off and they coasted to the shoulder. So now I have to find them a tow. No one wants to tow EV's. Finally found one company to do it. 8-mile trip to the charging station in Tomah. $1,000! Normal vehicle on the flatbed would've been $150.

So now we're at the Tesla superchargers. Guess what. Can't open the fine charging port because the battery is dead!!! The charging ports open, you guessed it, electronically!!! And we also can't open the doors now (had to close the one open door when it was loaded onto the wrecker). The owner's manual is in the on-board computer, but the battery is dead.

I got the occupants to a store where they'd be warm while calling the rental company to figure out how to charge this POS, so I'm not sure of the outcome. I had to leave for a crash report. EV's may be the way, someday, but certainly not today!! *I'll stick with my dinosaur burner.*

Pedal Power:

Yes, it is possible to generate electricity by pedaling a bike connected to a generator. This concept is known as human-powered electricity generation or pedal power. The principle behind this is converting mechanical energy generated by pedaling into electrical energy using a generator.

The electricity generated can then be used to power various devices or stored in batteries for future use.

Pedal power has been used in various applications such as charging small electronic devices, lighting up homes in remote areas, and even powering concerts and events.

It is a sustainable and eco-friendly way of generating electricity, as it does not rely on any external source of energy, such as fossil fuels.

However, it is important to note that the amount of electricity generated will depend on several factors such as the efficiency of the generator, the strength and endurance of the person pedaling, and the duration of pedaling.

Generating enough electricity to power large appliances or industrial machinery may require significant effort and time. Additionally, it is essential to ensure safety measures are in place when setting up a human-powered generator, particularly when dealing with high voltages.

The cost-effectiveness of electricity generated by non-fossil fuel sources depends on various factors such as the initial investment, maintenance costs, and the cost of fuel or energy required for generating power.

Generally, renewable energy sources such as solar, wind, geothermal, and hydroelectric are becoming increasingly cost-effective as technological advancements are reducing the initial investment and maintenance costs.

Needless to say, the cost of refueling this generation source might make this type of renewable on the very expensive side...

I bet Gold's Gym or Planet Fitness would attract customers if these were in their equipment line-up.

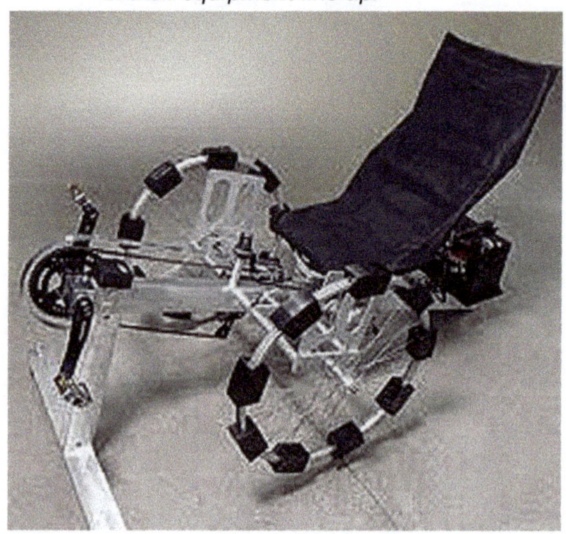

However, it is essential to note that the cost-effectiveness of electricity generated by non-fossil fuel sources can vary depending on the specific location, climate, and available natural resources.

For instance,
while solar power may be more economical in areas with plenty of sunlight, wind power may be more profitable in areas with high wind speeds. Therefore, it is crucial to evaluate the specific costs and benefits associated with different renewable energy sources before making any investments.

Ready Kilowatt, a popular mascot for municipal electric companies in the mid-twentieth century, may not be able to become a superhero for electrification, but the concept of electricity itself has the potential to transform communities and even entire countries. In many parts of the world, access to reliable and affordable electricity is still a luxury, and millions of people rely on traditional, polluting sources of energy such as coal, oil, and wood for lighting, cooking, and heating.

By expanding access to clean and renewable sources of energy, we can improve health outcomes, create new economic opportunities, and reduce the harmful environmental impacts associated with fossil fuels. From small-scale solar installations in rural villages to large offshore wind farms powering entire cities, electrification offers a pathway towards a more sustainable and equitable future.

However, achieving universal electrification will require a concerted effort from governments, civil society, and the private sector. This effort will involve overcoming numerous challenges such as inadequate infrastructure, limited financing, and political barriers, as well as ensuring that energy systems are designed to meet the needs and priorities of local communities. Nevertheless, by working together, we can leverage the power of electricity to shape a better world for ourselves and future generations.

Do you think L A Fitness should install these in all their Gyms!!!

ELECTRIC PLANES

It's a Bird,

No,, it's a Plane,

Nooo,,, its an Electron!

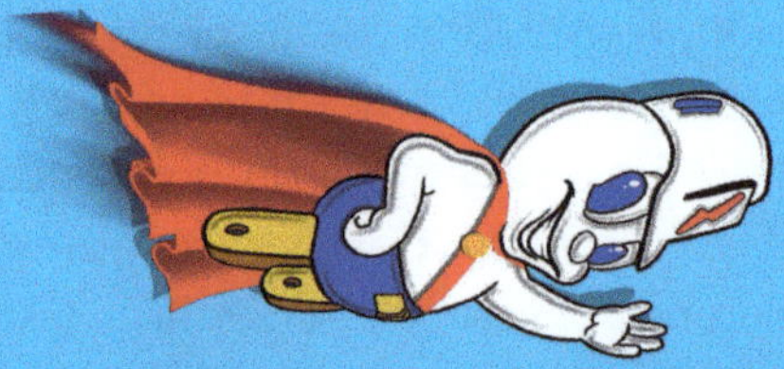

N337EE
AMPAIRE
N302AX
A6-RTA

Private Jets:

Private Aviation today represents 2.5% of the world's carbon emissions (and likely much more in non-carbon emissions) yet only 1% of the world's population are responsible for about 50% of all aviation emissions. Moreover, private planes are up to 14 times more polluting, per individual, than commercial planes and 50 times more polluting than trains, according to reporting by Transport and the Environment.

So **going to an electric powered plane is, I guess, a natural progression** that we need to think about and develop. Afterall, planes do create a great deal of CO_2' especially those jumbo jets but, the heavy weight of those batteries has been a nagging problem. And the weight does not decrease as the flight goes on ~~ batteries' always weight the same at take-off as they do on landing~~~ (Eddie Current is weightless).

Today, more than 250 electric powered plane companies are currently developing worldwide. Some expert observers predict that that electric airplanes will be commonplace before the end of the next decade. You can always find the 'talking heads' that will verify everything you are trying to use to confirm beliefs.

Consider as a major significance the number of parts and component manufacturers. In the U S alone, more than 1400 of these support manufacturing businesses total the aviation industry.

Motivation includes: higher efficiency, quieter, safer, and, much GREENER! (much greener than aircraft relying only on traditional internal combustion engines). Lets understand that electricity already plays a very critical role in todays aviation. Electric motors drive many components necessary for control of the aircraft, and the avionics provide seriously critical functions in navigation, convenience and safety.

Electric motors were invented in the 1830's, and battery-powered cars were first invented and manufactured in the 1890's, however the next huge changes in aviation will not take hundreds of years but will happen in the next decade.

Following are just a few of those that have already developed and are currently filing for certification on various design strategies and models.

As we discovered in the previous chapter, mobility is one of the hottest sectors, with start-ups and traditional OEMs constantly developing new technologies and transportation options. ***e*-Planes** are certainly a sector that is gaining a lot of attention from investment companies and one of the disruptive trends and technologies that will shape the future of mobility and the impact that they will have worldwide.

Tesla Just Silently
Filed Paperwork
That Could Change Everything,
I mean Elon not Nickola ~~~

Now, why didn't I think of this, a hybrid plane,,,

A **hybrid electric plane** is believed to have set a new record for hybrid aircraft endurance during its flight over California this past December. The Electric EEL from Ampaire, an aircraft systems company, covered 1,375 miles and flew for a total of 12 hours, with more than 2 hours of fuel and battery reserves remaining after its flight.

Drones can be used for knowledge acquisition,,,

Lets talk about Drones. They are small, light weight, need only short periods of flight, don't need to carry heavy payloads and have most of the critical navigation technology already developed.

Think about miniature Eddie Currents (electrons). Are these electrons smaller, Lighter in weight, than their brothers that drive the huge diesel electric train engines?

Students can learn about various drone operation technologies, including aerodynamics, aviation, programming, and data analysis, by working with drones. Drones serve as a tangible tool that encourages exploration and problem-solving.

Drones are basically Flying or floating Computers powered by Eddie Current.

Military Drone Functions & Modern Warfare Impact,,,

These unmanned aerial vehicles have revolutionized warfare through their applications in intelligence gathering, target acquisition, precision strikes, force protection, surveillance and reconnaissance, search and rescue, logistics and supply, and combat operations.

Today, they are fossil-fuel powered, however, who know what programs are in the works

Drones are basically Flying or floating Computers powered by Eddie Current.

Lilium Jet(a California based firm) has developed the first vertical take-off and landing jet. Its proprietary technology is termed **Ducted Electric Vectored Thrust.**

Jet engines are integrated into the wing flaps and provide advantages in aerodynamic efficiency, payload, and a lower noise profile.

It is now entering a phase of component testing. The first conforming aircraft is expected in May and battery testing is also scheduled, ahead of conforming aircraft production in Q3/4 2023. Lilium declines to discuss delivery timelines but confirms EASA certification flight trials will begin soon and certification is expected in 2025.

The Lilium Jet uses motors that develop 100kW. The maintenance data we have from automotive experience suggests the only time you'd need to remove the motor is after a bird strike or similar incident. It will fly forever.

What's a Whack???
How can I be out of it???

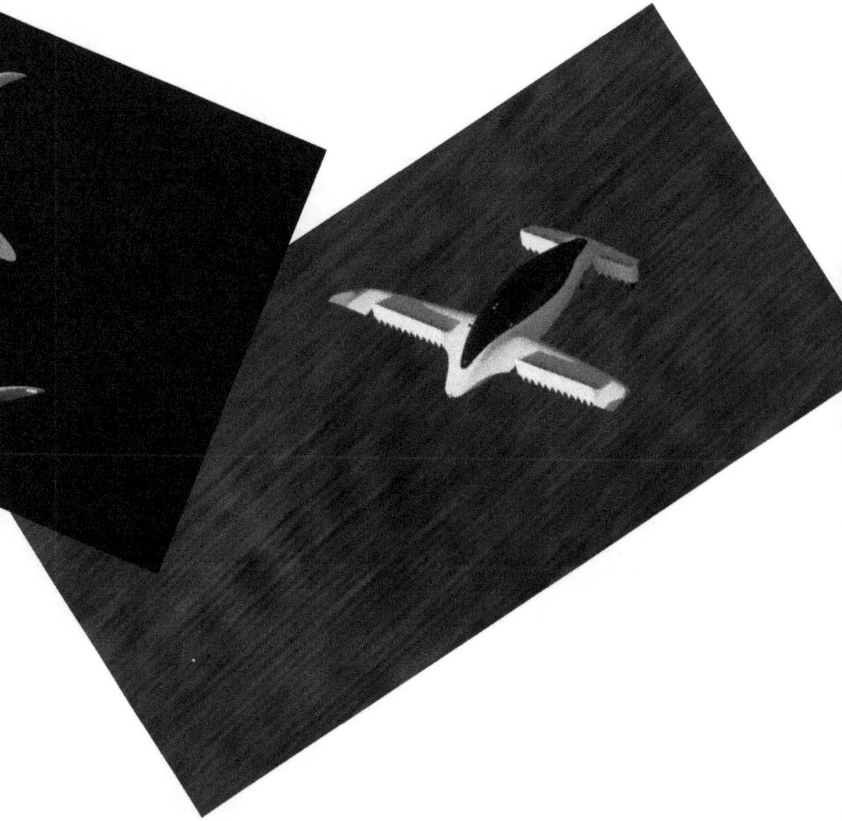

Vertical lift and cruise power comes from multiple electric motors driving ducted fans. Each unit is effectively a jet engine. A turbine, or small-diameter fan, forces air through a narrowing duct, where it is compressed and exits via a variable nozzle. Mounted in clusters on hinged wing and canard surfaces, the jets move as one in the transition from vertical to horizontal flight.

Aircraft design is all about compromise. The helicopter's vertical takeoff and landing (VTOL) capability, for example, comes at the cost of a draggy speed and range limiting main rotor. There is also the irrefutable fact that taking off vertically requires huge amounts of energy.

Considering contemporary battery energy density, an electrically-powered VTOL aircraft is further compromised, since high energy consumption in vertical flight strictly limits its range. Lilium's eVTOL concept, called Lilium Jet, aims to optimize the compromise.

Ducted fans or propellers?

VTOL, Vertical Take-Off and Landing, has emerged as a major criteria for this new category of electric powered planes, but why do they need 6 motors and propellers when a standard helicopter has only 1 huge propeller?

JOBY is another venture headquartered on the west coast that is developing an electric powered aircraft to function as an air taxi in major cities. Powered by six electric motors, the aircraft takes off and lands vertically, giving the flexibility to serve almost any community.

Today, a team of more than 1500 passionate engineers, experts, and leaders, are all focused on bringing this pioneering vision to life.

Headquartered in a world-class manufacturing facility in Marina, CA, Joby also has offices and workshops in Santa Cruz, San Carlos, Washington, D.C., and Munich, Germany.

AutoFlight's Prosperity eVTOL air taxi flew autonomously between the Chinese cities of Shen-zhen and Zhuhai.

This Chinese version utilizes separate propellers for lift and forward propulsion.

AutoFlight is designed as a five-seat air taxi—created for a pilot plus up to four passengers. Its recent flight, in the Hong Kong area, was fully au-tonomous.

ELECTRICITY STORAGE

'EverReady Eddie' is a close relative to Eddie Current

Much like refrigerators enabled food to be stored for days or weeks so it didn't have to be consumed immediately or thrown away, energy storage lets individuals and communities access electricity when they need it most—like during outages, or when the sun isn't shining or the wind is not strong enough to turn the giant blades. Storage can reduce demand for electricity from inefficient, polluting plants that are often located in low-income and marginalized communities. Storage can also help smooth out demand, avoiding price spikes for electricity customers.

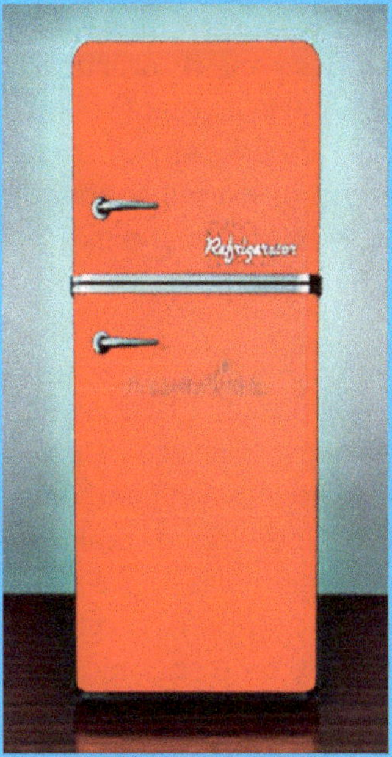

Pumped Hydroelectric Storage

Battery storage of electricity

Where does this chapter begin???

Ah, non battery storage of electricity. Now, there's the age-old debate about electricity and its practical uses. While batteries can certainly be a great tool for storing electricity, many people are unaware of the other, less conventional methods of storing energy. So, why not have some fun with this topic and explore the humorous possibilities of non battery storage of electricity?!

Let's start with the most obvious option: Charging up chickens - yes, it is exactly what it sounds like! We all know that chickens love to scavenge around and peck at things, so why not harness their energy by making them charge electric cars with their beaks? It's a simple concept - just attach the car's output port to the chicken's neck and make sure they keep pecking away. You'll soon see your car's energy level increase in no time!

If chickens aren't quite your thing, then what about using the force of a river or a stream to generate electricity? Picture it: attaching powerful turbines to a body of flowing water and watching as electric current is produced from it. What makes this even funnier is that adding more water flow will only increase the rate of electricity generated, which means you can shout 'Give me more power!' to the river - and it just might respond.

Finally, let's not forget the wonders of wind power - who said you need to attach turbines to a particular building to reap the benefits? Instead, why not use individual kites and hang them off of tall buildings? When the wind blows, it will cause the kite to lift, thus engaging the turbine to rotate and generate electricity. It's an ingenious way of harvesting the energy of the wind while also having some fun with colorful kites in the sky.

Ultimately, non battery storage of electricity can bring many unexpected surprises - and laughs! Unfortunately, lithium-ion batteries are not the only way to store electricity. While they remain the most popular choice, there are a number of alternatives that offer different advantages.

For instance, one approach is to use chemical reactions. These reactions create saltwater solutions, which can then be stored in a tank or other container. Over time, the liquid will naturally break down into its basic components, releasing energy in the process.

Another option is to use flywheels; these are mechanisms designed to store energy in the form of motion. Generally speaking, the flywheel is attached to an electric motor, which then stores rotational energy. This works great for applications that require frequent bursts of power, as the flywheel can quickly provide it.

In addition, some researchers are experimenting with using the power of smell to store electricity. Specifically, specialized sensors can detect the smell given off by flowers and convert it into electrical energy.

Let's Start This Discussion with Lithium

When the world's most valuable lithium company announced plans last year for a $1.3 billion processing plant in South Carolina, local officials hailed the project as transformative for the Palmetto State.

As someone once said...
"Is this the rest of the story?"

For a deeper look, you might want to check out the page titled "Disaster Capitalism." The high-tech project from Charlotte, North Carolina–based Albemarle was designed to process multiple sources of lithium—including recycled batteries—and supply this critical mineral to South Carolina's rapidly growing electric-vehicle industry.

Less than a year later, however, those plans have been slowed by a dramatic crash in battery-metal prices, driven in part by slower-than-expected growth in electric-vehicle sales in both the United States and China.

As a result, Albemarle has deferred spending on the project. The decision is part of broader companywide cost-cutting efforts that include layoffs and delays to other investments.

Producers of lithium and nickel, both key materials used in lithium-ion batteries for electric vehicles, have also been forced to stall projects and even close mines in order to conserve cash after a sudden and painful collapse in commodity prices.
Prices for lithium have fallen as much as 90 percent since the beginning of last year, while the price of nickel has dropped by roughly half.
The rapid boom in battery materials—once fueled by predictions of explosive EV growth—has suddenly collided with the realities of global markets, supply chains, and shifting demand.

*Please refer to **Disaster Capitalism** on page 74*

So much for this background info.

Now, let's start at the beginning on the next page,,,

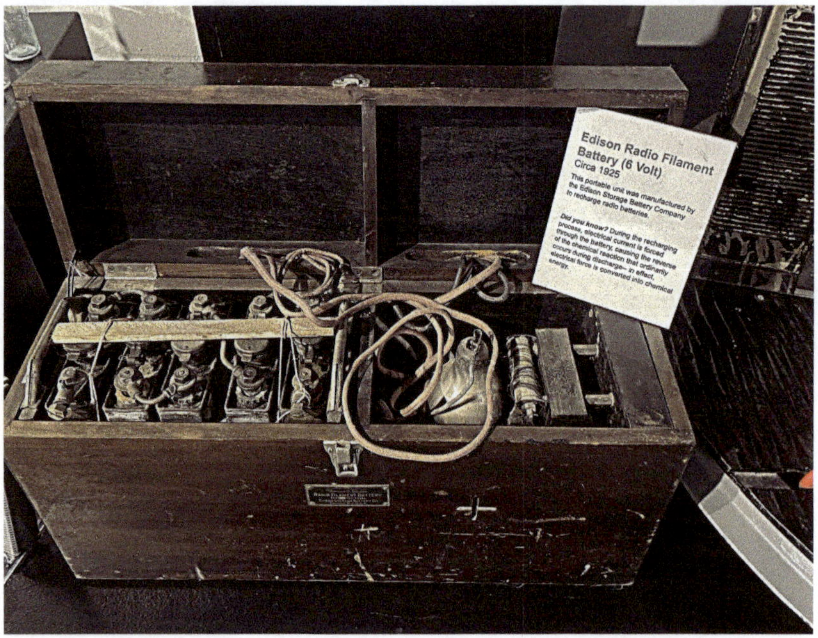

60 years on nickel hydrate-iron oxide batteries! My golf cart batteries don't last that long...

Did all this start with Thomas Edison???

From the winter estate of Thomas Edison in Fort Myers, Fl...

A view of the Edison lab and workshop in Florida.

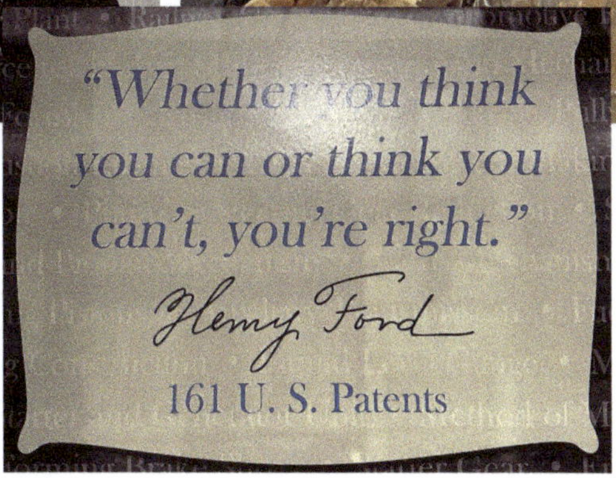

Edison Nickel-Iron Alkaline Battery

Circa 1910

In 1899 Edison began conducting extensive research on secondary batteries in hopes of finding a viable alternative to the existing lead-acid batteries which were heavy, difficult to recharge and corroded quickly. His idea was to produce a new and improved storage battery, one that would be strong and stable enough to propel an electric vehicle. Edison's research with alkaline cells led to the eventual development of a Nickel-Iron-Alkaline storage battery that he claimed would "run for 100 miles or more without recharging." His new batteries were designed to be used for industrial purposes such as electric delivery trucks, submarines, mining lamps, as well as train lighting and signaling. The Edison Storage Battery Company became one of the most successful divisions within Thomas A. Edison Inc. during the 1920s and 1930s.

Besides generating electricity, a giant balloon filled with liquid air could have several other practical uses. For example, this technology could potentially be used for cooling or heating applications. Liquid air could be stored and later released or circulated into buildings to help maintain comfortable temperatures.

In addition, this type of system could generate renewable energy in remote areas that are not connected to the electrical grid. Energy produced by solar panels, wind turbines, and other renewable sources could be stored and released when needed.

Such systems could also act as energy buffers, storing excess power generated during periods of high production and releasing it during times of higher demand. In this way, liquid-air storage could help stabilize energy supply.

Last but not least, systems like this could help reduce emissions, since the stored energy can be generated and released without burning fossil fuels.

Yes, it is entirely possible that large-scale systems using liquid air storage could help power parts of a city. In this approach, stored energy is released and converted into electricity when demand increases. That electricity can then supply homes, businesses, and public infrastructure.

With continued advances in engineering and storage technology, the process is becoming more efficient and increasingly cost-effective. What may once have sounded like a far-fetched idea is now being explored as a realistic option for supporting renewable energy systems and balancing future power grids

Liquid air batteries and gravity storage are among a number of solutions.

Liquid air energy storage is a fascinating topic and one with significant potential for use in modern energy systems.

One humorous way to imagine this technology is to picture a giant balloon used to store energy. In this playful example, the balloon is filled with liquid air and pressurized, allowing it to store large amounts of energy. When the balloon is "tapped," some of the liquid air evaporates, expands, and can be used to generate electricity. Pretty cool, right? Well... maybe not exactly in balloon form—but the underlying concept is surprisingly real.

We have all watched wind and solar power evolve from exotic technologies into common pieces of engineering, now competing with fossil fuels in many situations. Fossil fuels still retain an advantage primarily because of their reliability and consistency of supply.

To overcome this reliability gap, the energy industry needs large-scale, affordable grid storage capable of matching the scale of electricity generation. Today, the dominant technology for this purpose is the lithium-ion battery, the workhorse behind everything from mobile phones to electric vehicles. These batteries are reliable, scalable, and well-understood.

Many proposed alternatives, however, are clumsy, poorly understood, difficult to scale—or all three. But there is one intriguing possibility that might eventually challenge lithium-ion batteries: liquid air energy storage.

At approximately –196°C, the gases that make up air liquefy. Producing liquid air is a routine industrial process powered by electricity. Once liquefied, the air occupies only about 1/700th of the volume it would at room temperature.

When this liquid air is warmed and allowed to expand back into gas, it expands forcefully.
Using a device known as a Dearman engine—named after its inventor, British engineer Peter Dearman—that expansion can drive turbines and generators, allowing part of the energy originally used to liquefy the air to be recovered as electricity.

At present, a pilot plant in England recovers just over half of the electricity used in the liquefaction process. With further engineering improvements, the efficiency is expected to increase.

One promising improvement involves capturing the heat generated during the air-compression stage, possibly storing it in materials such as oil or molten salt. Reusing that heat during the expansion phase could significantly improve efficiency and bring the system's performance closer to that of large lithium-ion battery installations

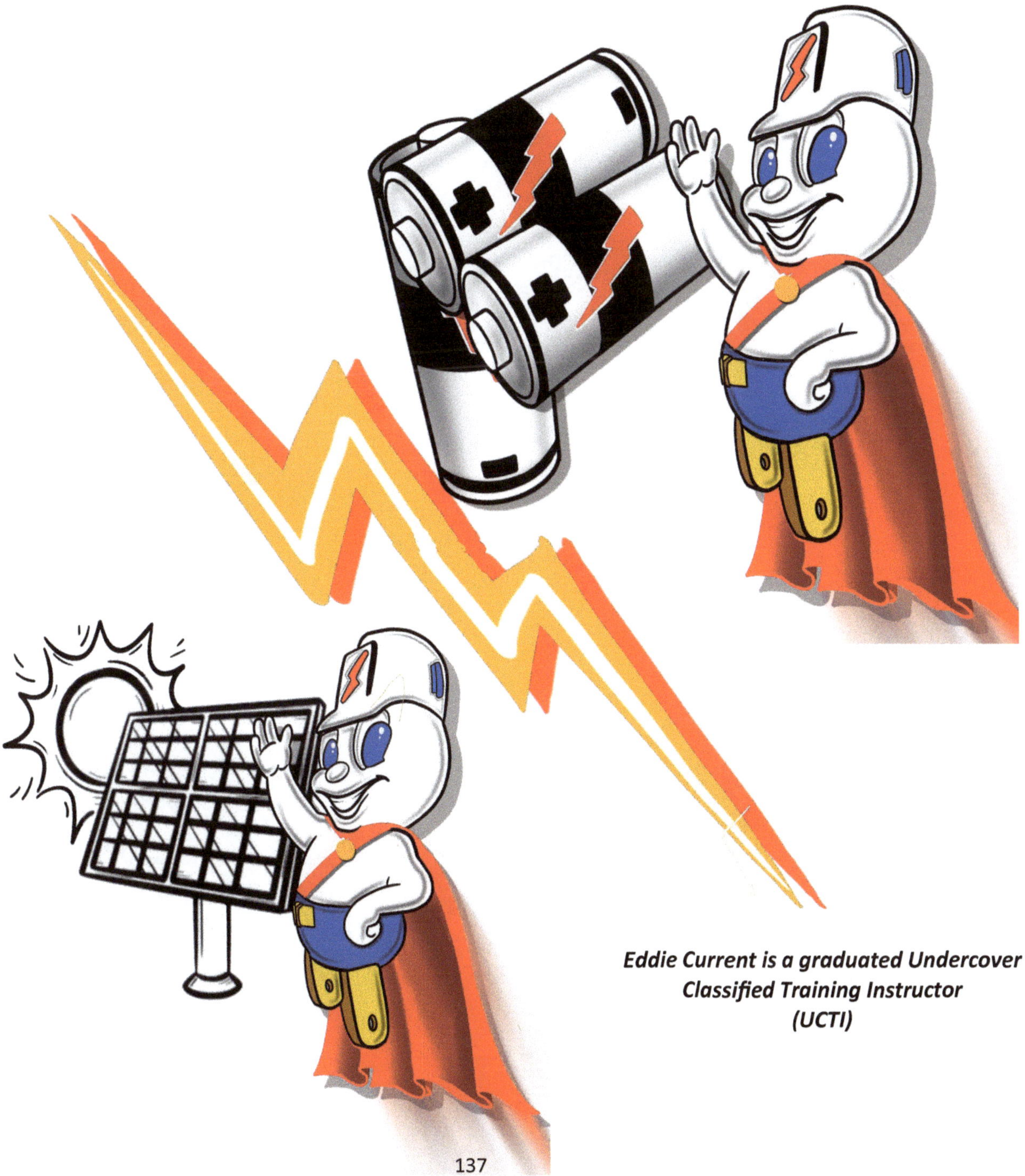

Eddie Current is a graduated Undercover Classified Training Instructor (UCTI)

Hybrid Power Plant:

Here is a comprehensive package that integrates (engine power plant and energy storage) seamlessly with renewable generation, ensuring stable and uninterrupted power generation, resulting in both economic and environmental benefits.

"My wife said to me that the spark between us had gone. So, I tasered her, and I'll ask her again when she wakes up."

By attaching a turbine to a kite, you can generate electricity simply by letting the kite soar in the sky.

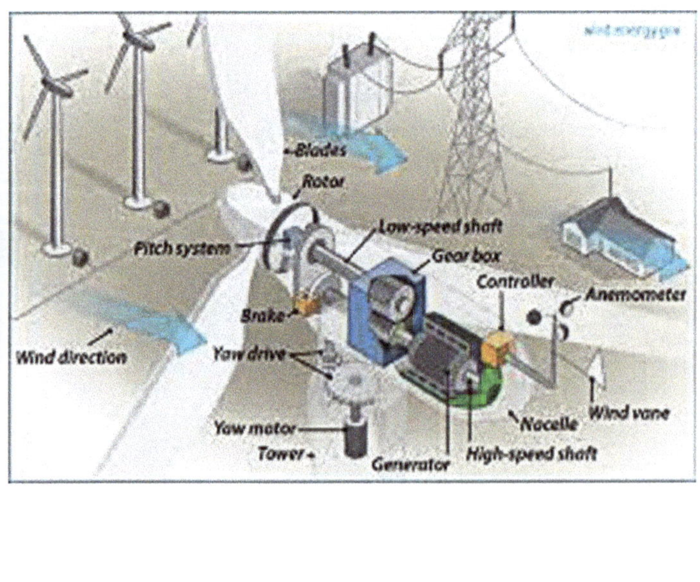

Compressed Air Energy Storage (CAES)

Compressed Air Energy Storage (CAES) is a method of storing energy by compressing air and holding it in tanks, underground caverns, or other storage spaces until the energy is needed. When electricity demand increases, the compressed air is released, expanded through turbines, and used to generate electricity.

This technology is considered promising because it allows excess energy—often produced by renewable sources such as wind or solar—to be stored rather than wasted.

There is also a slightly amusing side to CAES. The same principle of compressed air can be used to power mechanical systems, including things like roller coasters and amusement park rides, which adds a bit of entertainment value to the concept of energy storage. While grid-scale systems are far larger and more complex, the underlying idea remains similar.

Another advantage of CAES is that the air itself is continuously reused. The same air can be compressed, stored, expanded, and recycled multiple times, making the system both sustainable and relatively cost-effective.

In addition, CAES systems can provide important grid services such as frequency regulation and black-start capability—the ability to restart portions of the power grid after a blackout. These capabilities make CAES particularly attractive to utilities seeking reliable backup and balancing power.

Because the compression process can be powered by renewable energy sources, the overall system can contribute to a cleaner and more sustainable energy infrastructure.

Overall, compressed air energy storage represents an efficient and flexible method of storing energy, helping stabilize power systems and support the increasing use of renewable energy.

Could CAES power something as ambitious as a rocket to Mars? In theory, perhaps—but in practice, it is not the most suitable option.

Rockets require extremely large and consistent energy outputs, and CAES systems are better suited for grid-scale storage and short-term power balancing rather than continuous propulsion.

For space travel, more reliable and energy-dense storage systems—such as chemical fuels or cryogenic propellants—are far more practical. These systems provide sustained power over long periods and are lighter relative to the energy they deliver, which is crucial for spacecraft efficiency.

In other words, while compressed air can help power cities, stabilize electrical grids, and maybe even make roller coasters more exciting, it probably won't be launching astronauts to Mars anytime soon.

SGES Large-scale energy storage technology (SGES) Large-scale energy storage technologies are essential for maintaining stability in power systems that rely heavily on renewable energy. They also play an important role in addressing both the global energy crisis and environmental concerns.

One emerging solution is Solid Gravity Energy Storage (SGES), a mechanical energy storage technology designed for large-scale applications. The basic concept is simple: energy is stored by lifting heavy masses and then recovered when gravity pulls those masses back down, turning generators in the process.

Research on SGES has grown rapidly over the past decade. A number of studies between 2010 and 2021 have examined the evolution of this technology and explored its future potential. Investigations have focused on equipment design, operating principles, materials, system performance, and mathematical modeling.

Compared with other large-scale storage technologies, SGES offers several important advantages, including:

- High cycle efficiency (typically 80–90%)
- Large energy storage capacity
- Good geographic adaptability

Relatively low cost.

Gravity-based energy storage systems build upon a simple physical principle that has been used for centuries. In fact, some of the oldest energy-storage systems in the world already rely on gravity. Hydroelectric dams store energy by holding water at higher elevations, and tidal power systems harness the natural rise and fall of ocean waters.

Modern gravity storage technologies expand on these ideas by using large weights, towers, underground shafts, or other engineered systems to store energy and release it when electricity demand increases.

Traditional gravity-based storage can also be found in familiar structures such as water reservoirs, elevated tanks, and even old stone quarries, which can be repurposed for energy storage.

One company developing this concept reports that Gravity Power systems can return energy to the grid at approximately 4¢ per kWh, less than half the cost of lithium-ion batteries—even after accounting for energy lost during the storage cycle.

As renewable energy continues to expand worldwide, gravity-based systems such as SGES may play an increasingly important role in balancing the grid and storing excess power generated by wind and solar resources.

A Gravity Battery is a type of electricity storage system that stores energy in the form of gravitational potential energy.

In a gravity battery, energy is stored by raising a heavy object to a higher elevation. The higher the object is lifted, the more potential energy it stores. When electricity is needed, the mass is lowered, and the stored gravitational energy is converted into electricity through a generator.

Gravity batteries typically use excess electricity—often produced by renewable sources such as wind or solar power—to raise the mass. When the mass is lowered later, the system generates electricity that can be returned to the grid.

One simple example involves lowering a large weight, such as a block of concrete, which turns a generator as it descends.

The most widely used form of gravity-based energy storage today is pumped-storage hydroelectricity. In these systems, water is pumped to a higher elevation using surplus electricity. When energy is needed, the water is released downhill through turbines, generating electricity.
The earliest devices that used gravity to produce controlled mechanical motion date back centuries. One example is the pendulum clock, invented in 1656, which used gravity and an escapement mechanism to regulate its movement.

Since then, gravity-based energy systems have evolved into sophisticated technologies capable of storing electricity on a large grid scale.

Pumped-storage hydroelectric systems were first developed in Switzerland in 1907. The concept later spread to the United States, where the Connecticut Electric and Power Company introduced the technology in 1930.

Since that time, pumped-storage hydroelectricity has grown into the largest form of grid-scale energy storage in the world.

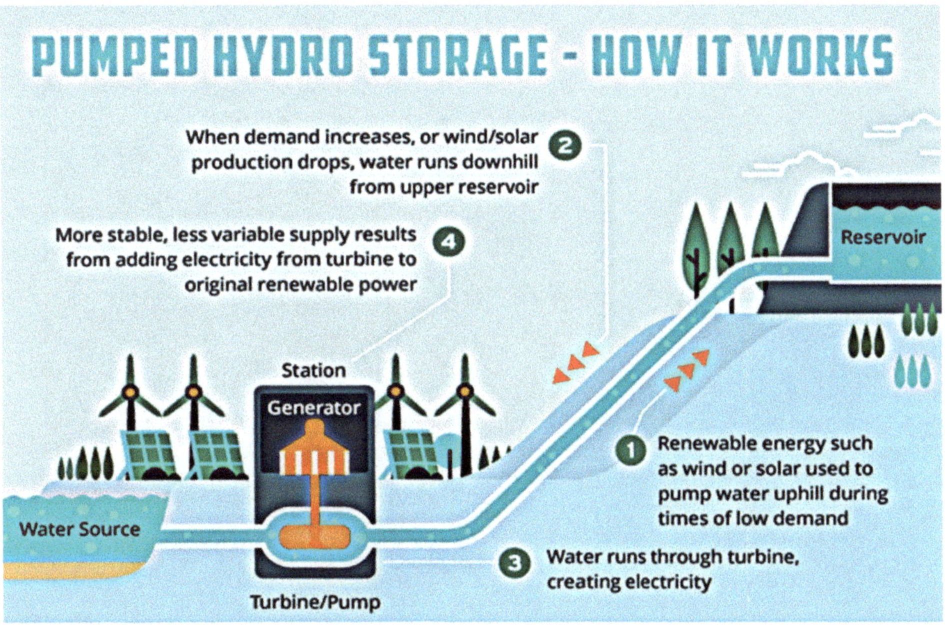

https://www.ucsusa.org/resources/how-energy-storage-works#:~:text=Flywheel%20Energy%20Storage%20Systems%20convert,(up%20to%2060%2C000%20rpm)

ENERGY ALTERNATES

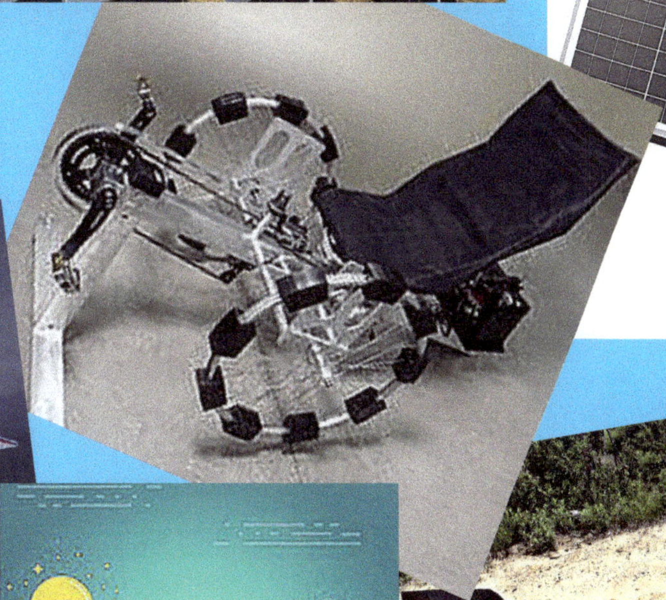

That last chapter might have bored a few of you guys, Now, here is a page you can share with your friends and impress your kids.

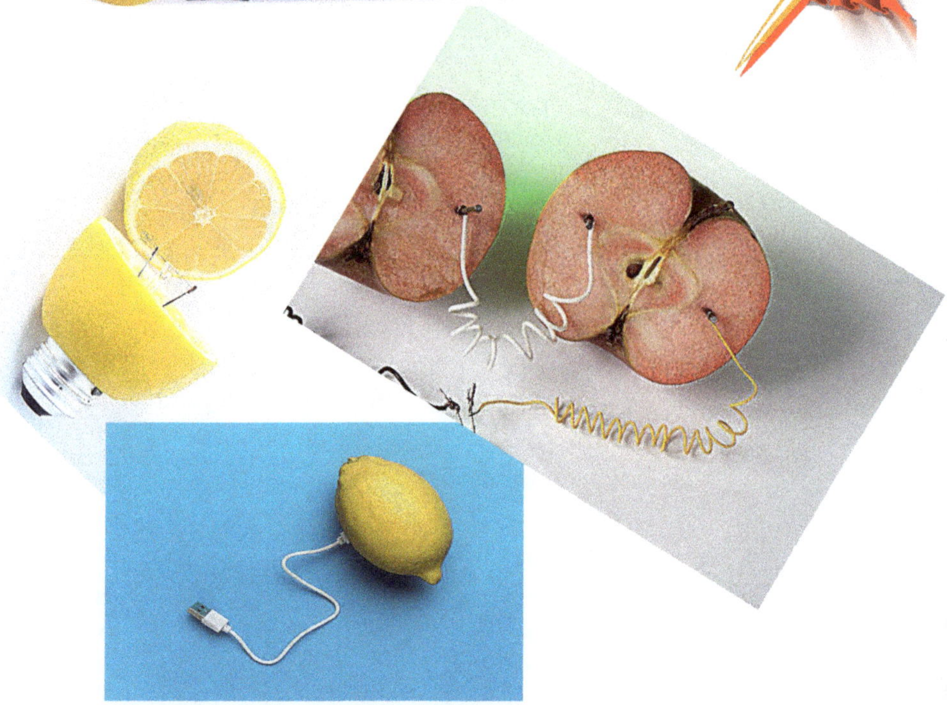

The Citric Acid in lemon juice is what makes the lemon taste sour, and acids contain ions which conduct electricity. The ions in acids are part of the chemical reaction that helps to produce electricity.

When you insert a piece of zinc metal (such as a galvanized nail) and a piece of copper (such as a penny) are inserted into a lemon and the wires are then connected, power is generated by reaction of the metals. This power is sufficient enough to be used to power a small device such as a light-emitting diode (LED).

By connecting a wire between the two electrodes, electrons are free to flow from the zinc towards the copper electrodes and create a current as well as voltage difference.

Surprisingly, a potato can produce approximately 1.0 volts while a lemon can produce approximately 0.7 volts.

Another use of the potato is to use the juice and turn it into vodka... *Bet you didn't think you could get culinary lessons here also.*

~~~Watch out for the wind turbine blades! ~~~
~~~

One of the most hilarious ways to generate electricity using the force of gravity would be to build a massive swing set. This swing set could have ropes connected to large counterweights on either side. As people swing back and forth, their motion would cause the counterweights to rotate. That rotation could then be connected to a generator, producing electricity. It would be both amusing to watch and a surprisingly useful renewable energy source.

Another playful idea would be to construct wind turbines shaped like animals or popular cartoon characters. As the wind blows, the turbines would spin, producing electricity while also creating a visually entertaining display.

Finally, imagine installing an upside-down water slide designed to capture kinetic energy. As someone slides down, their motion could drive a small generator, turning the excitement of the ride into electrical power.

In this way, energy generation could become not only practical but also entertaining—proving that sometimes renewable energy can be fun as well as functional.

Many creative ideas have been proposed to make the process of generating electricity more fun.

One idea is to incorporate art into wind turbines. This could involve painting the turbines or adding sculptures and lamps to them, transforming them from utilitarian objects to beautiful works of art.

Another way to make the process of generating electricity more fun would be to create virtual reality (VR) simulations that teach users about renewable energy. These simulations could allow people to build their own wind turbines, solar panels, and hydroelectric dams, and explore the science behind how these systems generate electricity. Additionally, games and activities can be created to help educate people on the various forms of renewable energy generation. Finally, interactive art installations can be constructed using renewable energy technology, such as solar powered sculptures that light up when it gets dark. Such installations would both be visually stimulating and informative at the same time.

The Claim: Climate change is a 'scam' because Australia's trees and mangroves absorb more CO_2 than the country produces,,,

A July 15 Instagram post purports that one person's explanation of Australia's net carbon emissions proves that climate change is not real.

"Man gives crazy theory that proves climate change is a scam," reads the post, which includes a video of a person talking.

"According to government figures, Australia is producing 499 million metric tons of carbon dioxide per year," the speaker says, according to a text overlay on the muted video. "But what I also found is that one mature tree will absorb 48 pounds, or 21.77 kilograms, of carbon dioxide per year. And Australia currently has 24 billion standard size trees, absorbing 453 million metric tons of carbon dioxide per year. And that's just trees – doesn't include any other plant life – and particularly doesn't include mangroves, and ***mangroves will absorb 50 times more carbon dioxide per year than a standard sized tree***.

(Imagine Florida with more than 1300 miles of coast and all those mangroves.)

USA TODAY July 27, 2023

The **Daintree in Australia**, is a national icon. It's home to the oldest tropical rainforest on earth, estimated to be about 130 million years old. It's also Australia's largest rainforest, incorporating an area roughly 1200 sq km, and supports a staggering selection of animal and bird life.

This historical footnote is significant.

First, in the past millions of years, Daintree has contributed trees, plants, and most probably dinosaurs. Which Mother Nature, together with Father Time, has converted into fossil fuels that we use today.

Second, the trees today are helping maintain energy independence by absorbing tons of CO_2 each year.

Third, this rich natural eco-friendly is home to a large colony of bees, which we have already seen as a potential new major renewable energy source.

Credit DR JEREMY BURGESS / SCIENCE PHOTO LIBRARY www.sciencephoto.com

How to turn your car alternator into a wind generator...

If you want your own wind powered generator, just check out Mother Earth News. There you'll find all you need is a vehicle alternator with a built-in voltage regulator, a fan and clutch assembly, some pipes, a tower or pole, some wire and a battery that you can get off your golf cart.

Parts is Parts... and will always have a place for the creative mind.

Who said the transition from fossil fuels to renewables would cost trillions and take decades???

MOTHER EARTH NEWS
THE ORIGINAL GUIDE TO LIVING WISELY

Which is the Killer

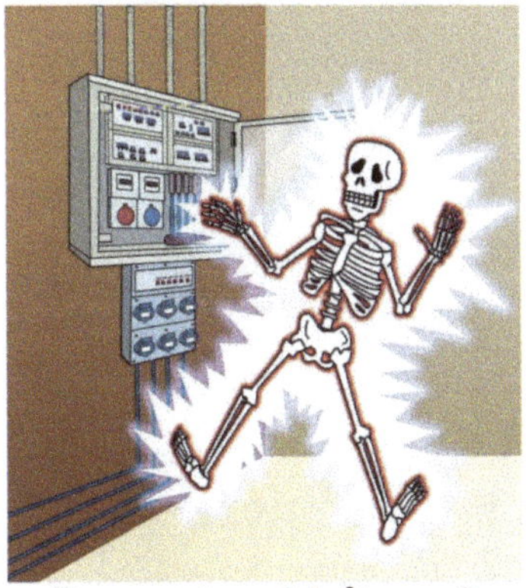

Current or Voltage?

Would this be an oxymoron???

"Let's design an electric chair with ultimate comfort as the main criteria…"

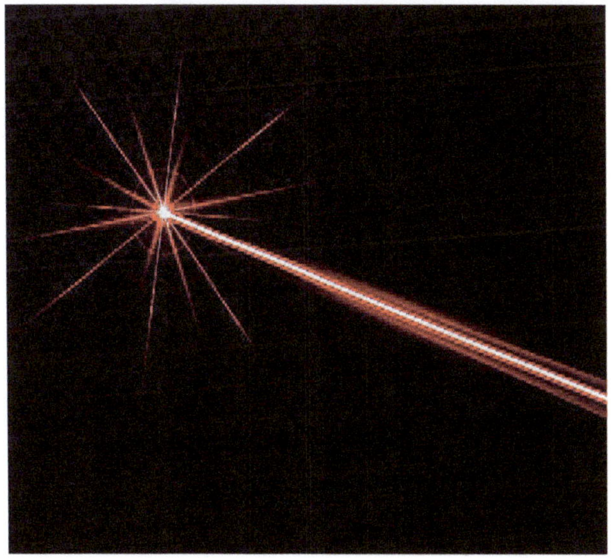

This could teach us that electricity can be a very powerful force, and we must never use it to hurt others, unless we need to teach them an important lesson.

Also, Consider the Laser:

A laser beam can be a delicate, precise instrument used in surgery. Its flight path is very straight and will not droop or sag.

Ponder this for a moment. 🤔
The wind farm in Mt. Pulaski has been running for 3 1/2 years. They have been replacing the generators in all the wind towers. There are 100 of them in this wind farm. So evidently the life span on the generators on these things is about 3 to 4 years. It takes 12 semi trucks and trailers, A 9 axle 500,000 pound crane, A 100,000 pound crane and 12 pick up trucks to change each generator. That is a huge amount of diesel fuel being used to maintain these wind towers. And the "Green Groups" would like You to believe they are all fueled by magic fairy dust.

What we never heart about is the **maintenance budget** of 'Going Green'... and with all these wind and solar farms, we are starting to realize the actual costs of the new green deal...

The US Transportation Department is investing $148.8 million repairing or updating nearly 4,500 electric vehicle charging ports in 20 states.

Pocket change compared to the $7 trillion already spent.

More than electrical

fantasy! Its really magic

fairy dust,,,

Electricity Generation

Electricity generation can be a very boring topic. But don't worry, I'm here to inject some humor into it!

Let's start by talking about renewable energy sources. Solar power is one of the cleanest energy sources around. It's like a giant sunbaked pizza on your roof! Wind turbines are also great for generating electricity. They look like giant pinwheels in the sky, and make a great addition to any landscape.

How about non-renewable energy sources? Nuclear power is one way to generate electricity, but it requires a great deal of safety as well as proper waste disposal. In other words, you have to keep an eye on your radioactive rods at all times! Likewise, burning fossil fuels isn't just bad for the environment. You could end up with a huge coal-fired chimney amidst the rolling hills of your backyard.

There are several new approaches to electricity generation currently being explored. One of the most promising is geothermal energy. This involves drilling down into the Earth's crust and using the natural heat found below to power turbines, or to create steam that drives generators.

Another exciting development is the use of ocean waves and tidal currents to generate electrical power. Wave energy has been around for some time, but recent advancements in technology are making it more feasible to capture this energy and turn it into an economical source of electricity.

Finally, research is being done on ways to use solar energy more efficiently, such as through photovoltaic cells and artificial photosynthesis. All of these technologies will help to reduce our dependence on fossil fuels in the future.

Yes, one type of solar energy technology that is showing the most promise for the future is photovoltaic (PV) cells. PV cells are made from semiconductor materials and they use the energy from sunlight to generate electricity. The advantages of this technology is that it is clean and renewable. It does not require any fuel or produce any emissions, making it an attractive option for reducing our dependence on fossil fuels.

Additionally, PV cells are becoming increasingly efficient at capturing the sun's energy, meaning that more electricity can be generated in a given space than ever before. Furthermore, advancements in PV cell construction have allowed for decreasing costs, thus making them more accessible to the public. Finally, PV cells have been integrated into many other technologies, such as vehicle charging and building-integrated photovoltaics (BIPV), further expanding the possible applications of this technology and increasing its potential impact on the future of energy.

PV cells are attractive for reducing dependence on fossil fuels because they do not require any fuel or produce any emissions. This means that by using PV cells we can generate electricity without any of the environmental damage associated with burning fossil fuels, such as air and water pollution. Furthermore, PV cells can provide a more secure source of energy since they are renewable, meaning that they can continue to produce energy indefinitely if properly maintained.

Another article, 7/31/23
https://www.benzinga.com/pressreleases/23/07/ab33460123/schneider-electric-and-pacific-gas-and-electric-company-announce-deployment-of-distributed-energy

https://www.theguardian.com/australia-news/2023/jul/31/australian-electric-vehicles-ev-sales-rise-increase

Compressed Air

Now, here's a theory that probably sprung up out of a pizza parlor in DC where conspiracy theories thrive!

Compressed Air Energy Storage **(CAES)** is a system that uses excess electricity to compress air and then store it, usually in an underground cavern. To produce electricity, the compressed air is released and used to drive a turbine. In a typical CAES design, the compressed air is used to run the compressor of a gas turbine, which saves about 2/3 of the energy needed to operate the turbine.

This leads to a reduction in natural gas consumption and can cut carbon dioxide emissions by 40 to 60 percent depending on the design. **CAES systems** have a large power rating, high storage capacity, and long lifetime. However, because CAES plants require an underground reservoir, there are limited suitable locations for them.

Only two commercial CAES plants exist in the world today, located in Germany and Alabama.

Compressed air energy storage (CAES) is also a form of energy storage that uses compressed air to store and retrieve energy. By compressing the air in tanks or cavities and storing it until it is needed, CAES can provide ancillary services such as frequency regulation and black start capabilities, making it an attractive option for utilities.

In addition, CAES can power roller coasters and other amusement park rides while still being sustainable. This is because when the stored air is released and converted back into electrical energy, the same air is recycled and reused, reducing the amount of energy wasted.

Furthermore, the process of compressing the air is often powered by renewable sources such as wind or solar, making the CAES process overall more sustainable than other forms of energy storage. Finally, its portability makes it an even more efficient and reliable source of energy. Overall, compressed air energy storage is an efficient and sustainable form of energy storage that can provide numerous benefits, including powering roller coasters and other amusement park rides.

Here's a project you can start in your backyard.

Scientists discover bees generate electricity!

Bees, those lovable buzzing creatures that fly around, are usually not credited with being able to make electricity. But, believe it or not, these industrious and busy little bugs are actually capable of doing just that!

This strange phenomenon has been studied by scientists who have discovered that bees can generate electricity through the process of *frictional induction*. When a bee moves along a surface, not only does it generate a small amount of static charge, but it also creates tiny electrical currents.

When a bee lands on a flower, the flower's petal provides a path for the current generated by the bee while it is pressing against the petal. As the bee moves along the petal, and more of its energy is transferred to the flower, the flower generates an electrical current. This current is what allows bees to get their food--nectar and pollen--from the flowers.

The stem then conducts the electrons into the ground where they are stored as we found out earlier.

So, how can all this electricity help us humans? Well, the most obvious use of the electricity created by the bees is the production of honey. Today we can see that the electricity created by these little creatures could bee strong enough to power not only their hives, but there could also bee enough left over to fire up a TV and recharge a smart phone.

This type of energy production is renewable, meaning it can be used without harm to the environment or depletion of resources.

So, the next time you see a bee buzzing around your yard, remember that it might be creating enough electricity to power your home some day! Who knows, maybe the future of electricity will depend on the tiny buzzing of bees!

This concept beats the idea of training bees to tap dance on solar panels for generating energy but it would require a tremendous amount of resources and effort to train bees for this task, as well as ensuring their safety and well-being.

If bees were in charge of the energy grid, there would be several key differences compared to our current energy infrastructure.

First, it would be much more sustainable. Bees generate electricity through photosynthesis, converting the energy from the sun into an electrical current. This type of energy production is renewable, meaning it can be used without harm to the environment or depletion of resources. It would also be cheaper than electricity generated from traditional sources such as coal and natural gas, since it requires much less energy to produce.

Second, it could lead to improved air quality. By producing electricity through a renewable source, we can reduce the amount of pollutants released into the atmosphere from burning fossil fuels. This will lead to improved air quality, resulting in fewer respiratory illnesses.

Yes, other renewable sources of energy, such as solar and wind power, are usually a much better option than the idea of training bees to tap dance on solar panels for generating energy. Renewable sources of energy such as solar and wind power usually offer much higher rates of energy production and require less cost and effort than training bees to tap dance on solar panels. In addition, renewable sources of energy are also much more reliable, with consistent and long term energy supply compared to training bees which can be unpredictable.

For example, solar power can be tapped during the day time when the sun is out, and wind power can be tapped during the night time when there is wind. Thus, these renewable sources of energy can provide a more consistent and reliable source of energy as compared to the idea of training bees.

No, unfortunately, it is not possible to get bees to generate enough energy to power a small city. Although bees are efficient and effective pollinators and have been used in some applications for generating energy, the level of energy production is still far too low to be able to power a city.

In addition, it would require a tremendous amount of resources and effort to train bees for this task, as well as ensuring their safety and well-being. Moreover, there are certain environmental risks with the idea of training bees to generate energy, as the fields and flowers used to feed the bees usually require pollination as well.

Today, renewable sources of energy such as solar and wind are much more feasible, effective and affordable options for powering a city compared to the idea of training bees, but, technology has a way of catching up once an intrapreneur gets involved!

Have you ever read that **"Wind is the cheapest energy source available?"** Now, please let me give you the rest of the story,,,

These public claims are based on cost estimates that assume the lifespan of wind turbines to be 30 years. However, according to the U.S. Energy Information Administration, the lifespan of wind turbines is about 20 to 25 years.

Actual data on installations is demonstrating far different data. MidAmerican Energy, which now operates more than 3400 in 37 different projects in Iowa, show wind turbines are reaching the end of their lives even faster. At one point they were forced to repower turbines merely 14 years after they were installed. Current histories of existing coal, nuclear, natural gas and hydroelectric plants are demonstrating that they can generate electricity for more than 50 years.

MidAmerican Energy has invested $14 billion in wind projects.

By not factoring in this additional spending, these reports not only underestimate the true cost of wind energy, but overestimate the cost of power plants capable of generating electricity for more than 30 years.

In other words, these cost-estimates tell us that wind energy would be the cheapest source of energy if all power sources produced electricity for a similar period of time ... but they don't!!!

Additionally, because wind turbines can only produce energy when the wind is blowing, they generate electricity less frequently than other generation sources. In Minnesota, wind farms produced electricity only 34.67% of the time in recent studies.

This is far lower than coal at 56.9%, hydropower at 64% and nuclear at 84.6%. Natural gas plants are utilized at a lower rate of 17.6%, but this is because natural gas are used as a "backup" source of electricity when the wind isn't blowing, (an additional cost that is not included in the cost of wind turbines.)

About MidAmerican Energy:

MidAmerican Energy, headquartered in Des Moines, Iowa, serves 813,000 electric customers in Iowa, Illinois and South Dakota, Information about MidAmerican Energy is available at midamericanenergy.com and company social media channels.

https://www.crowrivermedia.com/limited-lifespans-of-wind-turbines-result-in-higher-energy-costs/article_ce7a2c93-0e13-586c-aa8e-cbd8a826c76a.html

Wind farm photos that missed their day of glory on social media channels,,, As stated, the replacement rate is faster than expected and given the current very high recycling costs, there is a real danger that all components will go directly into landfill.

Wind energy is experiencing a boom, but in a pattern eerily reminiscent of the Pennsylvania oil boom in the early 1900's, wind farms are building ever larger turbines to farm wind energy further and further from shore. This trend carries risks, especially as turbines come with largely hidden costs. Increasing evidence suggests that although larger turbines can capture more energy, at a certain point the costs of maintaining and decommissioning large turbines located far offshore will outweigh the benefits of that energy capture. If wind farm operators are to avoid creating an environmental and economic disaster in the longer term, they need to begin factoring realistic maintenance and decommissioning costs into their projections.

The old, and the new

Here comes the storm!
Many people think lighting is harmful, but, Benny Franklin disproved this theory. Maybe this is an example of a power company in Tennessee recharging the Grid...

Electron Harvesting From Storms:

Here is your last option. In another new development, we will soon be seeing electron harvesting from snow.

Skiing enthusiasts have discovered that dust particles are electrically charged thereby attracting moisture. When this is measured in the winter, the cold air turns the moisture into snow and these flakes actually carry an electrical charge. Electricity forming in storm clouds comes about when the air is moving vertically up and down causing the moisture condensation to release its charge.

Water molecules align differently under electric fields and attach to dust particles at different temperatures. This is why a humid day does not produce any electrical charge, but storms with violent air movements do.

This is a seasonal phenomenon. In the summer, we have lighting, in the winter we have snow. One logical storage solution is easy in the winter—piles of snow piled up along side of the road or parking lot., or on the picnic table out on the veranda. In the summer, we have to be much more creative, like flying a kite.

It must be noted that noted that there is a fine balance between increased air pollution, dust and temperature that must be met by storms to produce snow.

And, once again, this source is not a reliable source of electricity.

Here is a recap of some of the data shared on previous pages... *Just what you really wanted,,,*

5 different types of energy storages

with efficiency from 25% to 90%

The energy sector has always been associated with engineering advancements. This will continue.

1. **Lithium-ion batteries**. "Fresh", agile, expensive Energy storage in lithium-ion batteries is considered one of the most efficient. But only until the battery begins to degrade.
   Pros: fast construction, almost instantaneous output of the stored energy (tenths of a second)
   Cons: price, degradation, absence of disposal, acquisition of rare earth elements Liion 88 – 90%

   One of the main problems with lithium-ion systems is that they degrade over time. The degree of degradation depends on the intensity of the battery use – how often it worked during peak loads, how often it was discharged to zero. But on average, after ten years, the battery capacity decreases to "economically disadvantageous" levels (it applies to industrial facilities) – 50 of the declared for new batteries. Less capacity is less efficiency. Degradation results another problem with lithium-ion batteries – their disposal.

   2. **Power-to-gas**. Gas storages: innovative, "green", less efficient
   Pros: Possible to supply methane to container transportation
   Cons: low efficiency, expensive electrolysis process
   Power-to-gas 50%

2. 3. **PSP Pumped storage power** plants are not as fresh as lithium-ion batteries or hydrogen storages.
   PSP(new) 75%
   Pros: inexpensive storage, high capacities
   Cons: expensive and lengthy construction, the need for a suitable landscape

   4. **Gravity energy storage** systems. Unusual, cheaper.
   The principle of work is based on gravity and friction, basically, it is similar to the PSP.
   Gravity storages 80 - 85%
   Pros: fast construction and comparative low cost
   Cons: new technology with uncertain efficiency

   Finally,
   5. **Thermal energy storage.** Or a heated "philosophical" stone by Siemens
   Pros: inexpensive
   Cons: Not very efficient, only suitable for seasonal energy storage
   Thermal storages 25%

https://kosatka.media/en/category/blog/news/5-sposobov-hraneniya-energii-i-naskolko-oni-effektivny#:~:text=Energy%20storage%20in%20lithium%2Dion,one%20of%20the%20most%20efficient.

Tesla - Toasters - Teslas

You have had this opportunity to see a few of the ways Eddie Current has enriched our lives and, a few of the media hypes on supposed abuses, most of which are deep dives into misinformation and hidden agendas. I am confident that you will see this publication as a contribution to point out complexity, involvement, and dependency on our need for power.

LASERS:

In the past and still today, entertainers use lasers for their colorful visual effects, and doctors use them to perform delicate operation, often to the human eyeball. Also, as electronic appliances, Lasers can be so powerful that they can vaporize a bulldozer 2000 yards away.

Instruction manuals had to be rewritten to include instructions to change the power setting from "Vaporize Bulldozer" to "Delicate-Eye Surgery" depending on the task at hand.

The END

Book 3

LET THE *SPARKS* FLY!

Adventures of Eddie Current

And His Wired Friends

A COMIC NOVEL

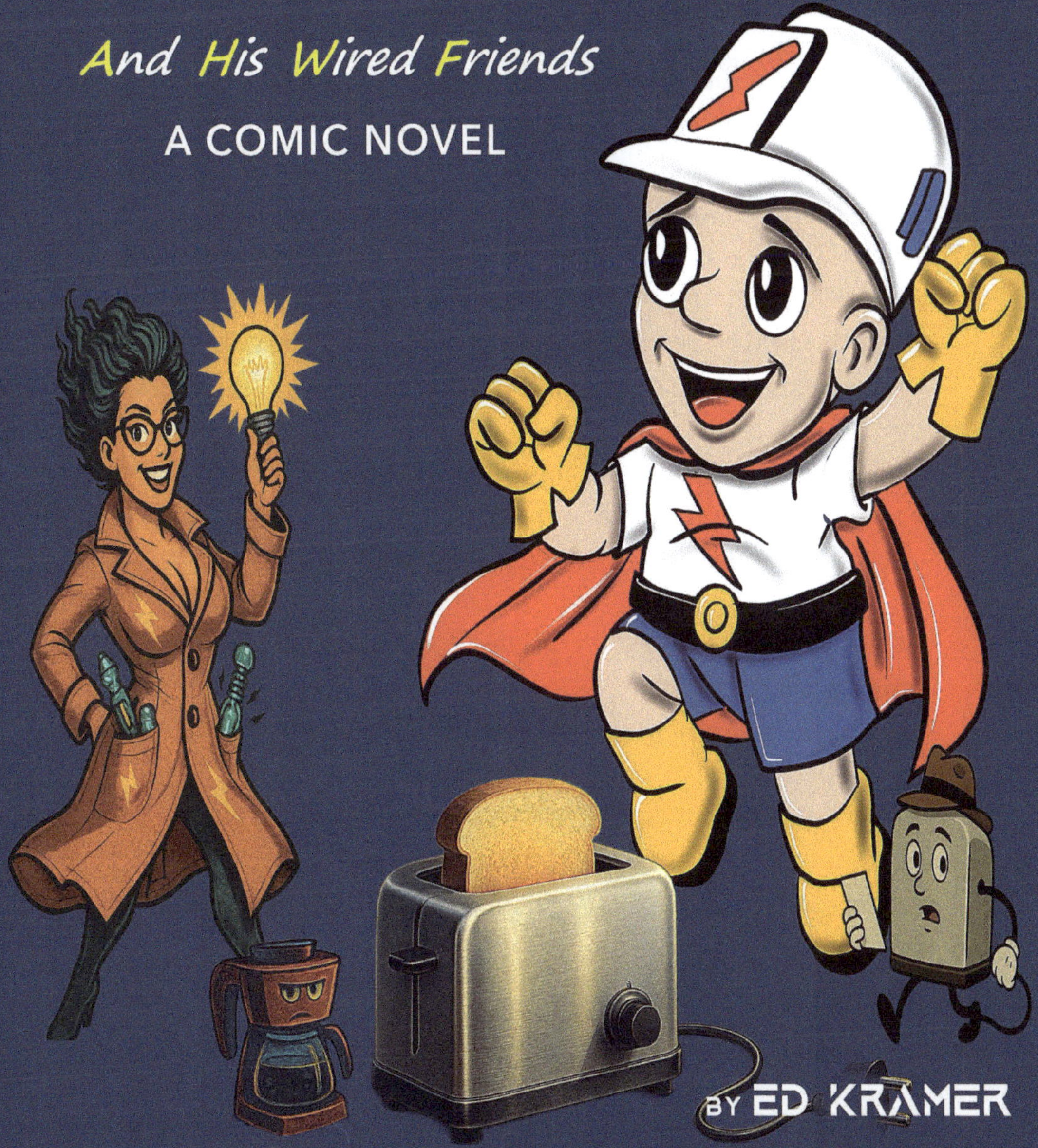

BY ED KRAMER

Recommendations from Contributors:

"Finally, a book that understands power isn't just about voltage — it's about *style*. I laughed, I learned, I recalibrated my surge suppressor."
— Mary Louise Watt, Power Theorist & Smart Grid Whisperer

"A shocking masterpiece! Every page had me tingling. I short circuited twice from the excitement. Five sparks out of five."
— Myron Volt, Retired Conductor & Casual Insulator Collector

"Eddie owes me a new wall socket, but this book? Brilliant. Gave me goosebumps on my grounding wire."
— Bob Transformer, Substation Saloon Regular

"I didn't *ask* to be included in this book. But now that I've read it, I suppose it's acceptable."
— Wendy Watt, Agent of the Renewable Rebellion

"Watt a ride! Eddie makes electricity less scary than my fourth grade science teacher — and he's way better dressed."
— Andrea Amp, Spunky, a little Impulsive, a Certified Live Wire

This following few pages were created with a little help from AI (which, you've got to know, is powered by the Power Grid anyway).

When chickens talk about climate change, do they use foul language???

Introduction:

We Don't See Electricity... But It Sees Us.

Electricity is everywhere. It's in our toasters, our toothbrushes, our televisions — even in that mysterious hum the fridge makes at 3 AM and yet, despite lighting our homes and powering our playlists, it remains mostly invisible... like a polite ghost that tidies up after itself and occasionally zaps your socks.

Scientists have told us that deep inside all this buzzing and blinking are electrons — tiny, energetic charged particles that make it all happen. But let's be honest, photons are a bit shy. You'll never see one smile, shake hands, or explain why your phone manages to die at 12%.

This story will grapple with:

The ethics of grounding your feelings,
* Whether light bulbs are really *sucking darkness,*
* Who is the Grid Team and who are its characters,
* Why do they keep giving me their emotional baggage,
* Whether the Substation Saloon is open 24/7,
* And an electrifying emissary named Eddie Current.

Eddie is your trusty tour conductor on this high-voltage journey through the world of electricity. Along the way, you'll meet his energetic allies and illuminating friends — like Bob Transformer, Breaker Bill, Charlie Capacitor and others, plus a few smart appliances with attitude.

This isn't just a science book. It's a whimsical, wire-tangled exploration of how electricity impacts our daily lives, our technology, and maybe even our moods (especially before coffee). or plug in, power up, and prepare to be shocked, enlightened, and slightly amused. There are NO sharp corners here so, let's flip the switch.

Eddie's ready. Are you? ***Warning:*** *Side effects may include gaining a magnetic personality, unexpected*

Power Grid Humor on the walls in the Substation Saloon

"If you plant a light bulb in your garden, does it grow into a power plant?"

Eddie Current's favorite flavor of ice cream is Shock-a-lot

If an ice making plant has a power failure, does it go into liquidation?

"A sweater I bought was picking up static electricity. So, I returned it and They gave me another one free of charge."

"People asked me how it feels when you stick your finger in an electrical outlet? To be honest, it Hertz!"

What Is Electricity — and where does it go when it leaves the toaster? **

Is this where we started? If you've ever stared at a blinking lightbulb, tangled with a rogue power strip, or wondered why your phone charger disappears each week, you're not alone. Electricity is all around us — zipping through wires, lighting up cities, and sneakily powering your smart fridge's gossip.

But don't worry — you've got a guide.

Meet **Eddie Current**, your voltage-savvy tour guide with a spark in his step and a shocking sense of humor. You will also meet more electrifying characters, including the watt-wise Mary Louise Watt, the ever-enthusiastic Andrea Amp, and the buzzworthy Myron Volt. Eddie is here to unravel the mysterious world of electrons, outlets, and everyday energy.

With vintage flair and modern mischief, this whimsically illustrated book dives deep into the world of Electricity — how it's made, where it travels, and how it secretly rules our lives. (Yes, even your toaster has opinions).

So grab your surge protector and hold onto your circuits — Eddie Current is flipping the switch on everything you thought you knew about power.

"Electricity isn't just watts and wires — it's life in motion, baby. I'm not here to shock you... well, actually, I am. Buckle up. This ride's got resistance."

*— Eddie Current, Ambassador of Power** This first book is considered to be a gradual, but not meteoric success.*

On the following pages you will hang out with **Eddie** and see graphic demonstrations of the *Theory Of Electricity.* This theory will be disclosed to you by insiders, an in depth, accurate view of each electron doing his or her job and how they spend their leisure time. You will get to meet the electrons that are responsible for the tasks that our lives depend on.

Eddie will be our guide into the mysteries of electricity. He began his career as an invisible electron, ***#e–127B,*** the Sizzler of Sourdough, and with experience, grew into his full-time responsibilities as **Ambassador of Power**.

One minute, he'll be racing through toaster coils with purpose. The next? He's questioning if he's AC, DC, or just a very energetic hallucination caused by a broken surge protector.

Joined by other emotionally repressed electrons like Myron Volt, a wiry, wide-eyed tinker who smells perpetually of ozone and adventure, Bob Transformer, who he really is, and a microwave wench named Wendy Watt and others, Eddie must navigate flickering lightbulbs, existential substation therapy, and karaoke night at The Substation Saloon — all while trying to answer the biggest question of all: Where does electricity go when it leaves the toaster, and does it leave its feelings behind?

You will also experience the real-life Shocking Adventures of Eddie Current & Friends, Wired Personalities, Life on the Line and finally, Amped: Chronicles from the Current Side. So let's go!

The Future: Life After Toast

Once upon a plug, in a small cozy kitchen, lived a brave bolt of electricity numbered ***#e–127B.***

Every morning at exactly 7:03 AM, ***#e–127B*** was jolted awake by the sound of a lever being pushed down and the faint, sleepy mumble of a human saying, *"Ugh... toast."* This was his cue. Showtime!

He zoomed down the copper highway, rocketing through the cord and into the heart of the toaster — the Gridiron of Breakfast. There, alongside thousands of his electron cousins, he hurled himself into the heating coils, sacrificing his energy to summon the almighty force of Heat.

Bread browned. Butter melted. A symphony of crackles echoed through the crumb tray.

As soon as the toast popped, his job was done. But he didn't

disappear — oh no. He zipped out of the toaster, full of post-toast adrenaline and a hint of smoke-scented swagger.

"Where to now?" he asked. "Microwave? Hair dryer? Maybe a power-hungry gaming PC?"

"Nah," said Bob Transformer, one of the Power Grid Team.

"We're headed to the Great Loop. Gotta return to the power station before we end the day at the Substation Saloon."

So off they went, surfing the neutral wire like it was an underground highway. They passed bored electrons snoozing in unused outlets, a rave happening in the fridge, and a dramatic soap opera unfolding inside the smart TV ("Will HDMI ever forgive USB-C?").

At the power plant, ***#e–127B*** clocked out and headed to the lounge. "Good shift?" asked Myron, sipping a high octane juice. "Toast again," Eddie shrugged. "You?"

Electric Blanket, "They still think I'm magic."

Then, suddenly, the alarm rang: **"COFFEE MAKER ONLINE! COFFEE MAKER ONLINE!"**

#e–127B, surrounded by fellow electrons, jumped up, tied on his lightning-bolt bandana, and grinned. "All right, boys — we ride again." And with that, he dove back into the wires, off to power the next miracle of morning survival or back home to do the chores of house-keeping...

So now you know just what it takes after coffee breaks between heroic missions.

Power Grid Humor on the walls in the Substation Saloon

THE INVENTION OF THE SHOVEL WAS A ***GROUND*** BREAKING DISCOVERY,,,

If you saw someone harassing electricity, would that be abuse of power?

But ,,, the INVENTION OF THE BROOM WAS THE ONE THAT TRULY SWEPT THE NATION...

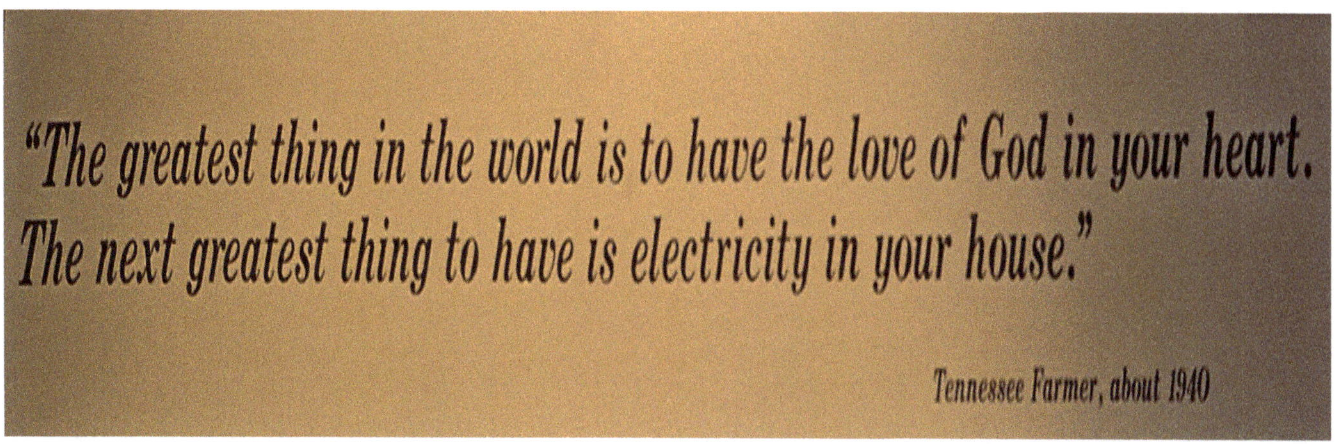

Eddie Current,

from a humble electron to

Ambassador of Power

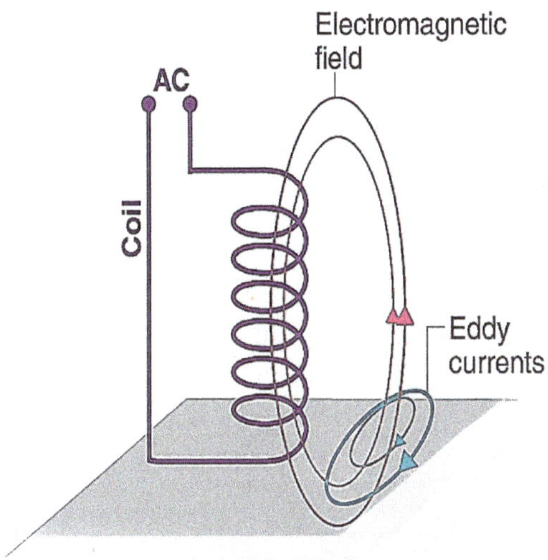

Was Eddie Current's name inspired by an Electromagnetic Phenomenon?

Soon you will be privy to the hidden truth behind electric power and how it affects our daily lives.

Eddie's mysterious activities will be explained and you will finally know the whole truth about his secret life... how he powers the alarm clock and where he goes after he finishes preparing breakfast.

Eddie with his first cape

#e–127B and his friends were always on the move in the bustling bustle of a copper wire. He was a high-strung, fidgety electron who couldn't sit still for even a ***Planck Second*** — *zipping* past protons, *zigzagging* through atoms, and *zapping* anyone who dared to call him "negative." What a nervous little zaplet he was! Always buzzing with anxiety about potential differences.

DEFINITION:

The term "Planck second" refers to a unit of time within the system of Planck units. It's the time it takes light to travel one Planck length in a vacuum and is approximately equal to 5.39 × 10^-44 seconds.

Power Grid Humor on the walls in the Substation Saloon

#e–127B struck a victorious pose. "No, my work isn't done. Somewhere, a coffee maker is crying. A gaming PC is dreaming of frame rates. And a curling iron just screamed, 'I NEED JUICE!'"

With a dramatic *snap* and a trail of sparks he vanished into the wires once more, leaving behind only the faint scent of ozone and the echo of his catchphrase:

**"Stay grounded, kids.
But never stop flowing."**

A now famous portrait with the toaster

Power Grid Humor
on the walls in the
Substation Saloon

Then, one fateful nanosecond in a hydroelectric power station, something *shifted*. A magnificent voltage surge swept through the conductors, and **#e–127B** was swept along with it. No longer just a jittery jumble in the circuit, he *began to flow* — elegantly, purposefully, electrically. He realized he wasn't just bouncing about in chaos any more — *he had become something greater. A current. A force. A movement.*

Up ahead, Mary Louise Watt, who was more than a Hydro-Hussie and backup vocalist for the band "The Ohm-an Empire," flowed out of the coffee shop powered entirely by passive-aggressive hydroelectrics.

She was one of the Power Grid crew, and she was radiant — literally. Her scarf was woven from recycled copper filament and her glasses sparkled with polarized photon filters. She was sipping something suspiciously luminescent.

When he saw her, **#e-127B** felt a jolt. It wasn't static. It was respect. Maybe raging hormones or a mild attraction. But mostly respect. She was smart, grounded, and watt-savvy. And he, just a youthful surge in flux, had just spilled half her thermo-electric mocha with his magnetic personality.

"So Sorry," he said, wiping the cup with his black electric tape. "Sometimes I overreact. Like a lithium battery at a drum circle."

She smiled. "You must be **#e–127B,** I've heard praise of your waveform. Not bad for someone who once got tangled in his own extension cord."

They talked. He sparked. She explained how energy flows are actually social organisms. He told her how he once got trapped in a dimmer switch for four days and had to learn to meditate.

One thing led to another, and soon they were talking of co-writing policy for Ampville's Grid Harmony Council.

Later that evening at a council meeting in the Substation Saloon, they met Bob Transformer (another of the Power Grid Team). Snacks of popcorn popped in a fondue pot and low-resistance cider being served.

As the party was reaching a crescendo, Bob Transformer (probably a little tipsy) stood up and declared:

"I NOMINATE **#e–127B** FOR
AMBASSADOR OF POWER!"

The air crackled with static. Someone dropped a popcorn resistor. Mary Louise adjusted her glasses.

"Actually," she said coolly, "he's already acting as a power ambassador every time he bridges a circuit with empathy and balanced amperage."

The room went silent. Then the jukebox (a refurbished Tesla coil) blasted "We Will Shock You" and the room erupted in cheers and applause.

Thus #e–127B became Eddie Current the acknowledged *Ambassador of Power.*

He didn't campaign. He didn't lobby. He had listened, charged wisely and occasionally danced like an overloaded Rhoomba robot vacuum.

And so, with a zap and a flair for the dramatic, he was given the title Ambassador of Power, Champion of the Flow, Lord of the Load. They gave him a shiny red cape made of hot ions and a hard hat. Now he was dressed for success. As he traveled the world — from toaster to turbine, from socket to starship — he spread the shocking truth: That even the most jittery, anxious electron can light up the world, if only he finds the right circuit.

Long live *Eddie Current!*

Eddie's T.O.E.

After years of surging through wires, dodging tripped breakers and philosophizing with wind turbines at sunset, **Eddie Current** unveiled his greatest insight: the **T.O.E.**, not the "Theory of Everything" (he left that to black holes and overcaffeinated physicists).

He introduced his **Theory of Electricity** as a comprehensive, slightly chaotic, and deeply illuminating framework which explains how power pulses through every corner of our lives. From the secret anxieties of smart appliances to the social behavior of electron's social media, his T.O.E. combines lived experience, live current, and an uncanny ability to monologue mid-surge. It's not peer-reviewed — but it *is* peer-energized.

Most folks think electricity is just numbers, volts, watts, amps, ohms and maybe the occasional lightning bolt if the sky's feeling dramatic. But after a lifetime zipping through wires, flipping switches, and once getting emotionally entangled with a smart blender, Eddie has discovered a deeper truth. A grand unifying principle. A theory so shocking, so enlightening, that, (*plug-in-and-buckle-up)*, it had to be called: T.O.E. — The Theory of Electricity. Not some chalkboard scrawl in a dusty lab — T.O.E. is *lived electricity.* It's about the pulse in a dance floor's 12 foot high subwoofers with 400 watts of power, the quiet hum behind your fridge's midnight thoughts, and the undeniable magnetism of extension cords that tangle no matter how neatly you coil them.

And he uncovered this theory alone, although he was guided, sparked, and occasionally fried by a brilliant cast of fellow electrons (the Power Grid Team fellows) each with a personality more charged than the last.

You've already met **Mary Louise Watt**, the watt-wise wonder woman with a brain like a circuit diagram and boots grounded in science. If electricity had a conscience, it would ask her first.

Next there was **Myron Volt**, the plaid-clad curmudgeon of conductivity, who's forgotten more about fuses than most have blown.

So we will follow the flow: from generation to transformation, resistance to resilience, and yes, with a few philosophical musings about dimmer switches.

In the end, you will understand that **Eddie's T. O. E.** isn't just about his foot or how electricity *works*. It's about how electricity *lives*.

Bob Transformer, who's seen all the highs and lows — and still shows up to the Substation Saloon in full step-down mode.

Andrea Amp, who is cooler, steadier. She wears a trench coat full of tools and has zero tolerance for inefficient wiring.

And let us not forget **Breaker Bill**, who's never wrong and never lets you forget it.

What a crew. Together, they've helped Eddie refine T. O. E. — not just as a set of ideas, but as a way of *feeling* the current of everyday life.

For us mortals the *Theory Of Electricity*

began with our human

Electric Pioneers

⇒ ***Benjamin Franklin,*** The Postmaster who spoke in proverbs.

⇒ ***Thomas Edison***, The man, the myth, the walking patent machine.

⇒ ***Nikola Tesla,*** The master of alternating current.

⇒ ***Luigi Galvani,*** An Italian scientist, physician, and part-time frog DJ.

The real story began one legendary afternoon in 1752 — the day **Benjamin Franklin** looked out of his window, saw a menacing thundercloud grumble overhead, and said, "What a shocking day for a stroll."

While the sensible citizens of Philadelphia slammed their shutters and dove under their quilts, old Ben reached for his most aerodynamic bit of string-and-stick and headed straight into the storm with the swagger of a man who thought, *"I wonder if this weather needs a kite?"*

With rain pelting and thunder rumbling like an angry sky bear, Franklin marched into the tempest with a twinkle in his eye and a key in his pocket. While others baked bread or organized socks, *he* was inventing science in the most theatrical way possible — by literally waving a lightning rod at the sky while holding on to it with a wet string. (Kids, don't try this at home or *ANYWHERE.)*

Thus, one soggy genius, one rebellious cloud, and one highly conductive house key danced their way into history, sparking not just lightning but the whole electrifying age to come.

This was a man who looked at a thunderstorm and thought, "*Weather this bad is definitely charged and needs a kite. I'm going out with my key."*

He harnessed lightning, invented bifocals, and once proposed that turkeys should be the national bird, for which the bald eagle never forgave him.

But still, he was unemployed. Despite inventing half the stuff in your kitchen drawer and revolutionizing science with a string and a key, he hadn't landed a steady gig.

Ben wasn't your average philosopher — inventor — scientist — statesman — diplomat -- humorist — electrician — almanac publisher or swimmer, he was the original multitool in breeches.

One minute he was charming the French with witty epigrams, the next he was electrocuting turkeys *on purpose*, then casually inventing bifocals because holding two pairs of glasses was "just too much fuss."

He didn't walk through life — he *waddled briskly*, quipping and experimenting with equal flair, all the while looking like he'd misplaced both his comb and his concern for lightning safety. Franklin wasn't just a Founding Father — he was *the Founding Father* who probably tried to patent the idea of Founding Fatherhood.

Then one day, the Founding Fathers called him into a candle-lit conference room.

"Ben," said George Washington, adjusting his powdered wig with two fingers, "you're on fire with these inventions and sayings. Seriously — A penny saved is a penny earned? I could print that saying on a beer mug or 'T' shirt and make a fortune."

Ben sipped his tea and replied, "Well done is better than well said."

John Adams leaned forward. "We've got an offer. You're the wisest guy we know. The most respected, the most unlikely to get distracted by fame."

Washington interrupted, "We want to put your face on money."

Ben blinked. "But I'm not a president."

"Exactly," said Adams. "We figured, who better to put on our biggest bill than a man who never ran for president, but managed to run the country half the time anyway?"

So, Ben's face was etched onto the $100 bill — the biggest flex in U.S. currency history. Hip-hop artists and rappers would later refer to these bills simply as "Bennies" or "Benjamins," which Benjamin Franklin would've absolutely approved of, though he ight've said, *"Flaunt not thy riches, lest thou attract both envy and a tax audit."*

But money wasn't enough. He needed purpose. A calling. Something *civil*, something *infrastructurey*... something to do with *mail.*

Once upon a time in colonial America — when wigs were high, shoes had buckles, and Wi-Fi was just a twinkle in the distant ether — he was a man with more ideas than a squirrel in a peanut factory.

Benjamin was so wise he didn't speak in sentences — he spoke in proverbs. For example, he'd say to his confused barber, "Early to bed and early to rise, makes a man healthy, wealthy, and... slightly socially unavailable" which may have been true, but was it tip enough for a poor barber with 5 children to feed at home?

In 1775, Thomas Jefferson asked Franklin to become the first Post Master General. "You want me to create and run the mail system?" he asked, scratching his chin with a quill.

"You're literally the only person who could get a letter from Boston to Philadelphia without it catching on fire or being eaten by a raccoon," Jefferson replied. And so he became he first Postmaster General — the original mail boss. He organized routes, standardized delivery, and made sure Aunt Martha's birthday card arrived *mostly* on time.

Rain, sleet, snow, bad handwriting — it didn't matter. Under Ben's watch, the mail got through. And whenever a new mail carrier joined, he'd offer a pearl of wisdom: "He who delivers swiftly, delivers twice."

Or, when the mail was late,

"The mailbox of life is always full, but it helps if you check it once in a while."

Franklin wasn't just a postmaster; he was the *proverbial* postmaster, literally stuffing the colonial consciousness with "early to bed" guilt-trips and unsolicited kite-based life insurance lessons. While others were fumbling with quills, Ben was zipping through the colonies on a mail route of enlightenment, armed with a satchel full of aphorisms and a key that could summon Zeus. Rain, sleet, or a lightning bolt — nothing stopped his delivery of common sense with uncharacteristically uncommon flair.

Epilogue: Benjamin Franklin went on to become a national icon. His inventions changed the world. His words filled almanacs, walls, and fortune cookies. His face adorned wallets. And his postal system, against all odds, actually worked.

He lived long, loved well, and died with a smile — and probably a proverb — on his lips. "Well done is better than well mailed, but both are pretty good."

Then along came **Thomas Edison**, the man, the myth, the walking patent machine. Often hailed as the greatest electrical pioneer he was definitely the most prolific.

Edison didn't just invent — he *multiplied* ideas. With over 1,000 patents to his name, he practically made innovation a full-time sport. He's best known for letting people read past sunset without setting the drapes on fire. But he also developed entire systems for electric power generation and distribution, helping cities light up like never before.

Of course, he wasn't without a flair for drama. One of his early inventions was the battery, leading him to be a staunch supporter of direct current (DC). This belief lead to a now-famous electrical beef with Nikola Tesla over alternating current — AC vs DC.

This so-called *"War of the Currents"* included everything from public demos to smear campaigns to one very unfortunate elephant. Still, love him or side-eye him, Edison's relentless drive, marketing savvy, and ability to *actually finish* his projects made him a towering figure in the electric age.

Without Edison, we might all still be fumbling with oil lamps and shouting, "Alexa, put a match to the wicks!" or "Siri, where is my box of matches?"

He was fueled by ambition and more than likely, a suspicious amount of coffee. Edison was frequently crowned the greatest electrical pioneer (at least according to his own press releases). He didn't just invent stuff, he patented everything that wasn't nailed down. Light bulbs? *Check.* Phonographs? *Check.* An electric machine called an Autopen (used to this day by presidents) that sort of worked but mostly looked like a prop from *Frankenstein*...
He checked that too!.

Thomas Edison is most famous for inventing the practical incandescent light bulb. But he didn't stop there. He built entire electrical systems, power plants, and even made electric chairs (we won't get into *that* awkward chapter). But, Edison was the DC (Direct Current) king — think of him as the original one-way street guy — and he was *not* amused when a certain tall, brooding, pigeon-loving wizard named Tesla came along with his fancy-shmancy alternating current (AC).

One of Edison's greatest achievements came in 1879 when he invented something truly revolutionary: ***The Electric Company.*** *Period.* That's right, not just a light bulb or a phonograph or a slightly alarming doll talking in his nightmares, but an entire *system* for delivering electricity to people's homes for an exorbitant amount of money.

The electric company is pure genius. It sends electricity to the customer through one wire, gets it back through another wire (because sharing is caring), and — here's the brilliant part — it sends that same electricity *right back to the customer again*. Over and over. Forever. It's like the world's most polite game of hot potato, but with volts. People were amazed. Lamps lit up. Toasters toasted.

Edison had done it. He made electricity not just a one-time discovery, but a utility — and he charged everyone monthly for the privilege. Pure genius!

Edison's relentless work ethic, his knack for self-promotion and ability to turn his lab into an invention assembly line, made him a powerhouse of the electric age.

Without him, you might still be reading by candlelight, yelling "scroll down" at a piece of parchment.

Thus began the **War of the Currents**, (NOT Currant War), aka *the most passive-aggressive science feud in history...*

These two brainy inventors didn't exactly send each other friendly birthday cards. Thomas Edison, the Wizard of Menlo Park, was Team Direct Current (DC) — he liked his electricity straight and serious. Then along came Nikola Tesla, the electric showman, with a wild new idea called Alternating Current (AC) that could travel way farther than DC could.

Edison wasn't a fan. So what did he do? He started electrocuting random animals in public to prove AC was dangerous. (RIP, unlucky goats and an unfortunate elephant). It was... a lot.

Nikola Tesla, on the other hand, didn't bother arguing — he just casually lit up the entire 1893 World's Fair in Chicago with his AC-powered light show like, "Ta-da!" People were amazed. He turned a whole city into a glowing disco ball before disco even existed. Spoiler alert: AC won. And today, thanks to Tesla, your toaster, smart watch, and tablet can now be charged wirelessly by placing the device on a device like an electric frypan. Your device is charged, with only a HUM. This is the kind of electricity — minus the drama — that we use today.

Part inventor, part wizard, part misunderstood space prophet, Tesla wasn't just ahead of his time — he threw a rave three centuries in the future and forgot to invite the rest of us.

Nikola Tesla. He was The Real Deal, The Master of alternating current — a rebel who didn't just think outside the box, he rewired it. He sent energy through the air, just for fun. They called him mad. "I call him *Dad," said Eddie.*

Nikola Tesla was one of the earliest and most brilliant pioneers of alternating current (AC) — the electricity that doesn't just go one way, it *vibrates* back and forth like it's dancing to its own techno playlist. Born in 1856 and possibly powered by coffee and cosmic energy alone, *Tesla* looked at the limits of direct current (DC) and basically said, "Hard pass."

While Edison was out there electrocuting elephants to prove DC's danger, (true story), *Tesla* was inventing AC systems that could power entire cities — and also maybe a death ray or two, depending on which of his notebooks you read.

He whipped up the AC motor, played with lightning like it was a party trick, and once tried to wirelessly light up the planet using a 187-foot tower and sheer audacity.

In 1780 **Luigi Galvani**, an Italian scientist, physician, and part-time frog surgeon, heard about Franklin's kite experiment, *(the kite that flew in a storm in 1752),* and stumbled upon one of the weirdest and most shocking (literally) discoveries in the history of electric science.

While tinkering with frog legs during an experiment that definitely wasn't approved by any ethics board, Galvani touched a bit of metal to the nerves of its legs — and *bam!* they kicked like they were auditioning for a dance contest.

The frogs were very much deceased, yet their limbs responded as if jolted by life itself. Galvani called it "animal electricity," suspecting that some kind of invisible lifeforce was conducting through the metal. In reality, he had unknowingly discovered bioelectricity and the basic principles of what we now understand as electrical impulses in the nervous system.

So, thanks to a curious man, a bit of metal, and some non-consenting amphibians, humanity took its first wild leap toward the science of electricity — with frog legs leading the charge.

What did he do with all this knowledge? His research suggested Human Behavior is caused by electrochemical processes and electrons running rampart through our bodies.

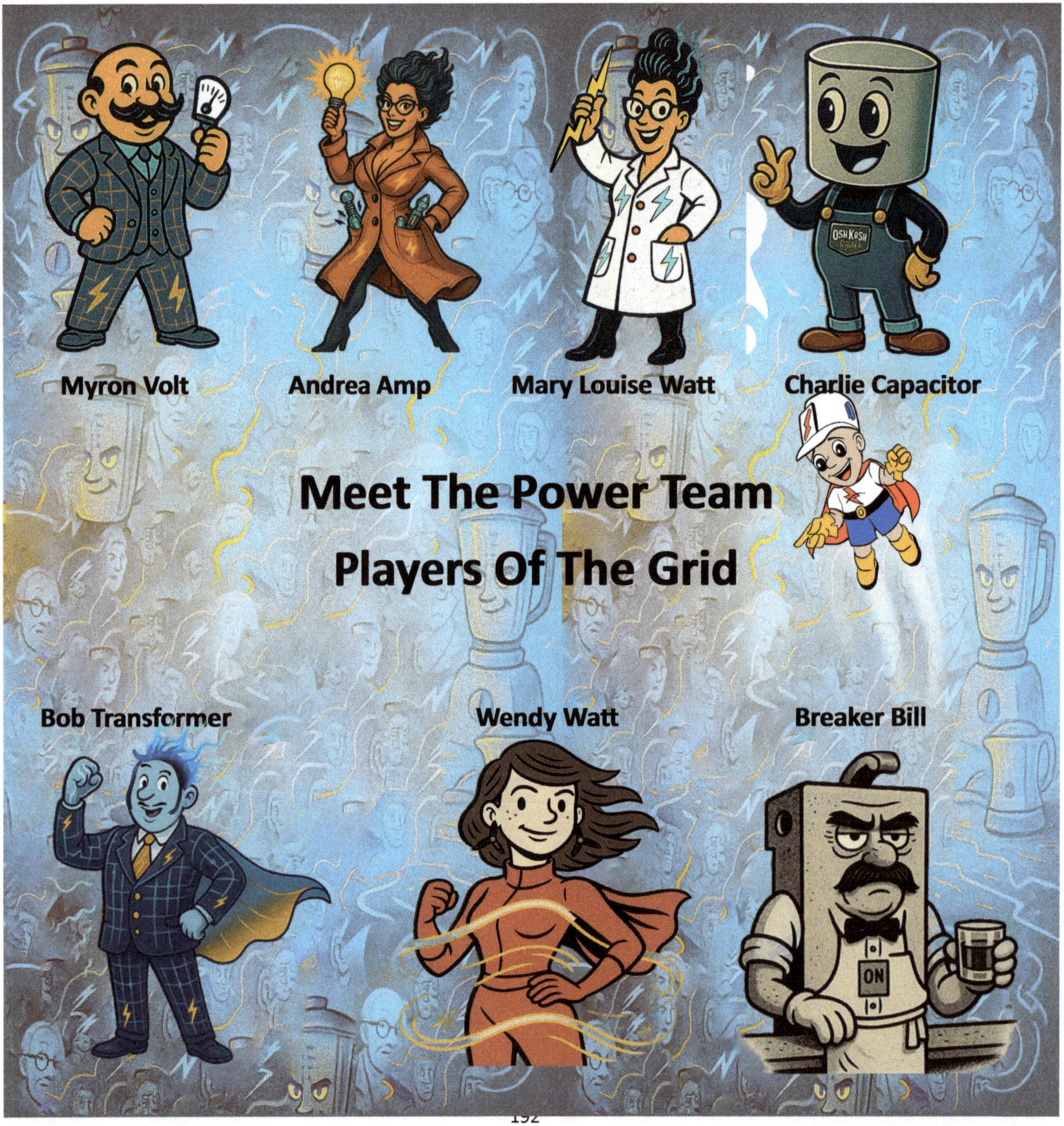
Myron Volt
Andrea Amp
Mary Louise Watt
Charlie Capacitor
Meet The Power Team
Players Of The Grid
Bob Transformer
Wendy Watt
Breaker Bill
ON

In the world of electricity, everything works because **Current, Voltage, and Resistance** know their roles—and they've assembled a team to make sure nothing goes dark.

At the center of the action is **Eddie Current**, the high-energy hero who keeps electrons moving where they're needed most. Eddie doesn't push or block—he **flows**, racing through wires with purpose and enthusiasm. Without him, nothing happens. With him, the grid comes alive.

Providing the push behind Eddie's motion is **Voltage**, represented by the seasoned and slightly crackly **Myron Volt**—a living spark who smells perpetually of ozone and wisdom. Myron supplies the pressure that gets electrons off the couch and into motion. When the system needs just the right amount of drive, Myron knows exactly how hard to push.

Balancing that push is **Resistance**, embodied by **Charlie Capacitor**, the steady hand of power. Charlie doesn't rush. He stores, smooths, and releases energy at just the right moment, preventing chaos and keeping the system calm. Resistance isn't the enemy—it's the reason everything doesn't burn out in a flash of glory.

Watching over the entire operation is **Mary Louise Watt**, a true hydro-powerhouse of a woman and the grid's guiding intellect. She measures the real work getting done, making sure power is delivered efficiently, responsibly, and with style.

When voltage needs adjusting, **Bob Transformer**, the step-down sage of the substation, quietly goes to work—reducing, increasing, and adapting power so it's safe and usable wherever it's headed. No drama. Just wisdom and coils.
Zipping in with modern flair is **Wendy Watt**, Agent of the Renewable Rebellion. Solar, wind, battery backup—you name it, Wendy's already on it. She brings optimism, speed, and a belief that the future is bright (and sustainably powered).

Then there's **Andrea Amp**, the smallest jolt with the biggest zap. She may be compact, but she measures the real intensity of the action—how much current is actually flowing. When Andrea says things are heating up, everyone listens.

And holding it all together—especially when things get a little overloaded—is **Breaker Bill**, the grizzled bartender at the Substation Saloon. Bill doesn't ask questions. If things get out of hand, he cuts the power, restores order, and pours a metaphorical drink until it's safe to try again.

Together, this Power Team doesn't just keep cities running, they keep **Current flowing, Voltage pushing, and Resistance protecting**—powering not only the grid, but the stories behind every jolt, zap, and flicker

In The Beginning

In reality, electricity doesn't start the day traveling to a distant land after leaving your toaster. Instead, it undergoes a transformation, turning itself into heat to crisp your bread. Once the task is complete, the flow of electricity halts until called upon again.

So, the next time you enjoy a slice of toast, remember the fascinating dance of electrons that made it possible. It's not just breakfast; it's a daily demonstration of physics in action and a requirement in the electron's operational manual.

We read earlier how *Eddie Current* started his career as a humble electron *#e–127B* working in the kitchen. His reputation as superhero of the socket, conductor of chaos, and sizzler of sourdough was achieved after years of practice and brilliant work.

With a dramatic *sparkle*, Eddie surged into action, racing through the wiring like a caffeinated rabbit in a lightning storm. He zipped through the toaster's coils, blasting raw, righteous energy into them and turning cold wires into red-hot breakfast warriors.

Bread browned. Crumbs burned. Toast which had once risen valiantly in the oven became, well, toast. But what happened *after* Eddie saved or served breakfast? He posed heroically at the outlet and gave a wink.

"They think I disappear when the toast pops. Nah. It's just intermission."

With a *zzzap*, Eddie jumped onto the neutral wire highway, back toward the grid.

But instead of heading straight home, he zipped into his favorite place: *The Substation Saloon*, a secret hangout inside a breaker box, lit entirely by leftover photons emerging from a Tupperware container with a cracked lid.

There the walls were lined with signed portraits of the greatest heroes of electric history. Eddie walked in, waved to the group of young electrons anxiously waiting for him to appear and began to speak.

"Gather 'round, kids. It's story time. Let me tell you of the legends of Ohms.

"I will begin by telling you of the Power Grid Team. First, I will speak of Myron Volt a powerhouse of a man. He is eccentric, occasionally moody, swinging from highs to lows. He was once electrocuted, but that only added to his magnetic personality."

Myron Volt is not so much a man as a living *spark* — a mischievous lightning bolt.

Without a hair on his head, and a volt meter always in his hand, Myron roams the world with pockets full of copper wire, half-scribbled blueprints, and dreams too big to fit in any one laboratory.

He lectures — truly, madly — that electricity is not just a force, but a *friend*, a shy creature that needs to be coaxed out with laughter, patience, and a good deal of singing to your tools.

Myron claims townsfolk often swear they saw him in the dead of night, conversing with his batteries or teaching currents to waltz through circuits as if coaching Ginger Rogers and Fred Astair.

They say that when Myron built his first generator, he did not test it with caution or care. He *invited* the lightning down from the sky with a brass umbrella, a wink, and a bow. The bolt that struck him didn't scorch him to a crisp (as many nervous villagers had feared), but instead infused him with a permanent, merry hum — and from that day on, every time Myron smiles, your hair will stand slightly on end.

He gave the world a thousand clever devices — some useful, some baffling, and some that simply blinked and beeped in polite confusion. But all bear his signature touch: a ticklish sort of magic, as if the very air around them wanted to dance.

As the years have gone by, you can still hear Myron Volt's laughter on stormy nights, tangled in the wires, chuckling through the outlets, a reminder that electricity, when treated kindly, loves to create a bit of good natured mischief.

Myron is also credited with the timeless expression:

"Theory of Electricity? — 120 or 121, whatever it takes." (In the rest of the world, it would translate to 240 or 241, whatever it takes.)

With that, Eddie and the youngsters moved on to meet Andrea Amp, who was at the saloon, tonight, in person.

All the electrons were thrilled.

Andrea Amp couldn't wait to put her two cents in to the discussion. She was sprightly, spunky, and — if we're being honest — a little impulsive.

Our pint-sized powerhouse Andrea lived in the golden glow of Gridville, tucked neatly between the Fuse Fairgrounds and the Substation Saloon.

She wasn't like the steady Volts or the sophisticated Ohms. She was an Amp — all energy, all the time. Her job? Moving electricity. Lots of it and Fast!

Clad in a snappy red jumpsuit with gold lightning bolts and pockets overfilled with highly charged tools, Andrea zipped around town, connecting power from here to there, there to somewhere, and somewhere to absolutely everywhere. Now was her time to decompress.

As she started to speak, the lights at the Substation Saloon began flickering. Andrea was sipping tea from a solar-warmed cup when she noticed them dimming.

"Myron," she said to the bald, plaid-suited genius beside her. "Either your eyebrows just absorbed all the lumens in the room or we've got a Darkness Sucker rebellion."

"I *did* just wax them," replied Myron. "But yes, something is definitely off."

Andrea replied, "Even though I'm an Amp, Myron, I think it's time to call in Bob Transformer."

Bob arrived in a blur, static crackling from his shoelaces. "What's up, Andrea?

Andrea handed him a Voltage Violation Meter and said "We think there's a rogue surge sneaking through the lines, or, it may be Darkness Suckers. Whatever it is, it's fast. Erratic… "

"That's full of beans, metaphorically, of course." Myron said with a chuckle. "It could also just be a mood swing from the neighbor's smart fridge. You can handle this. You go girl!"

With that, Andrea grinned, gave a nod and bolted. She surfed powerlines like a Pigeon on *Red Bovine*. She danced through outlets, interrogated a confused night-light, rode a bolt through two transformers and an emotionally lost Victrola record player.

Finally, at the corner of Spark & 4th, she found it. A miscalibrated hair dryer - plugged into a surge protector that had delusions of grandeur.

"HA! You again!" Andrea cried.

The hair dryer responded by blowing a defiant puff of heat and flipped every breaker on the block.

But Andrea was ready. She wrapped it in a grounding cord, whispered something soothing about wattage limits, and gently unplugged it.

Order was restored. Lights flickered back on. Somewhere in the back storage room, a smart fridge sighed in relief.

Back at the Substation Salon, Andrea strutted in to applause, carrying the subdued appliance like a firefighter saving an overly dramatic housecat.

Mary Louise smiled. "Nice work, Amp. You are shockingly reliable."

Myron chuckled, tipping his hat. "You might say this electron ***currently*** saved the day."

And Bob Transformer chimed in "I knew you could handle it."

Andrea grinned from ear to ear. She had proved yet again, to be the smallest jolt with the biggest zap!

Mary Louise Watt entered the Saloon, almost as if on cue. She was a powerhouse of a woman. Literally. She helped engineer hydroelectric systems in the early 1900's. When they said, "You can't," she said "Watch me generate 12,000 volts and change the world." And she did.

Mary Louise explained that electricity was not meant to frighten or bewilder — it was meant to comfort, to sing lullabies to sleeping houses, to trace dreams along the rafters at midnight. She talked about how her workshop — a cozy chaos of glowing jars, humming wires, and clocks that never quite agreed on the time — was known to pulse with a heartbeat of its own.

Mary Louise also known as 'The Bright Weaver', the 'Lantern Tamer', the quiet revolution tucked into a patchwork skirt wearing boots always covered with stardust brought the *glow* after the storm steady, soft, is powerful enough to guide lost travelers home.

Where others saw tangled wires and sparking chaos, Mary saw a secret language, a loom on which she could weave light into the fabric of everyday life.

She told the young electrons, "I was born on a morning when the mist curled itself into spirals and the first electric streetlamps flickered uncertainly in the dawn. As a girl, I would collect broken lanterns and dead batteries from the town's discard piles, speaking to them in a voice so gentle that even the oldest filaments seemed to tremble awake in my hands."

While Myron and the others courted storms, Mary courted the night, inventing delicate, clever ways to spread soft pools of light across kitchens, reading nooks, and bedside tables.

She told the group of young electrons again about her crowning creation, the *Kindle Loop* ** — a ribbon of golden current that twines through homes so sweetly that even the coldest, darkest winters seem to blush with warmth.

*** A situation where a Kindle device continuously restarts or repeats the same startup sequence, without fully loading into the operating system.*

Just at that moment, another Power Grid Card Carrying Crewmember entered.

"That's **Bob Transformer**," shouted Eddie.

"He's not just a name — he's a lifestyle with electrified hair. He takes high voltage and brings it down to earth, one step-down at a time."

Bob's greatest achievement was single-handedly stabilizing the town of Wattsonville during the Great Voltage Swing of '92 — when appliances were blowing fuses faster than you could say "brownout." He calmly stepped down the chaos and stepped up the coffee maker. Locals still say he saved the grid *and* breakfast that day.

Bob appeared, a small, glowing figure, a bright -eyed inventor with untamed hair.

In a voice sounding like he had just gargled with transformer oil, he said, "Mary Louise, tell us about the first time you met Myron Volt."

"Sure," Mary said. "I would love to." And with her eyes ionized, in an electric voice, Mary began, "I met him on an evening swirling with a mist that had a peculiar electricity — the sort that makes your ears hum and toes twitch in your boots."

"Myron was bounding down the crooked lanes of Windset Turbine Farm, his coat flapping like a sail, his pockets clinking with a thousand half-finished inventions. Behind him trailed a chaotic wake of sparks — little mischievous ones that leaped from puddle to puddle, sending the occasional sign into fits of flickering laughter."

"Meanwhile," Mary continued, "I was in a very quiet part of town, perched on the low stone wall of the square, where I was carefully binding silver filament into the frame of a lantern."

"I was whispering as I usually do — not a song exactly, but some soothing incantations to the sleepy wires to help them slide more smoothly into place."

Myron jumped into the conversation. "As fate would have it, one of my runaway sparks *pinged* off a can of Tomato Sauce, *bounced* off a security camera and *plopped* squarely into the lightning bolt Mary Louise was holding, which flared to life with a soft, golden *whoomph* "

"Yes" Mary said. "I dropped my pliers, untangled myself from all the wires, and looked up just in time to see a wild-eyed man skidding to a halt before me, his mustache sticking out in every possible direction as though he were the human version of a "static-charged cat."

"Oh!" Myron had gasped, clapping a hand to his chest. "You caught it!"

"I believe," Mary said, with the serene air of someone who has been catching wayward wonders all her life, "it caught *me*."

There was a long pause in the story, as both Mary and Myron became lost in their beautiful memory together of that night so long ago, when the space between them glowed steadily brighter, casting both their faces in a halo of soft, buzzing gold.

Then Myron had bowed so low he nearly fell, and introduced himself with a grand, tangled speech about coils and clouds and collaborative resonances.

Mary, in turn, had smiled a small, secret smile and simply offered him a plate of crackers and cheese from her small picnic basket — a gesture so kind and unexpected that Myron forgot every grandiose word he'd planned to say.

Here the story ended. But from that night on, the Substation was never quite the same. Some say the air in town began to purr with a softer kind of magic — a magical blend of Myron's wild, reckless sparks and Mary's patient, glowing threads. A magic that could light not just a home, but a heart.

And Eddie Current was keeping an eye on every activity.

Grace
JACK
ACKEREL
omato Sauce
OMEGA
3
NUTRIENTS
425 g (15 oz)

After that fateful evening when a stray spark and a golden lantern introduced Myron and Mary, the Substation Saloon began to buzz with a kind of secret glee — as though the very cobblestones were practicing a new dance.

But no great story ever grows without a third note to complete the chord. And that note — deep, rumbling, and slightly off-key — came in the form of **Bob Transformer**.

At this point, the entire crowd of young electrons turned to Bob, eager to hear his inspiring tale.

Bob was a mountain of a man, broad as a barn door and twice as sturdy, with hands like old tree stumps and a voice that could set large church windows rattling in their frames. Yet for all his size, Bob was a creature of kind, delicate thoughts. Yet for all of his size, Bob was a kind soul. He moved carefully, spoke thoughtfully, and tinkered with a tenderness that would make a baby bird blush.

Bob was unique. Bob didn't invent so much as *translate*. He listened to the growls of heavy machines, the sighs of worn-out dynamos, the grumbling of ancient water wheels.

He not only understood them, he empathized. He spoke the secret language of heavy things — the language of gears and magnets and forces that could topple trees if not properly coaxed.

Bob began: "I clearly remember the day I first met Mary and Myron. It was a frosty morning in the busy market square. Myron was running through the crowd, chasing a fast moving rental clock which refused to tick in any direction but backwards. Mary was off to the side, testing a new handwoven electric shawl she had just finished.

I was there too, not far from Mary, desperately trying to fix the mayor's lamplighter wagon which had recently begun belching blue smoke and whistling "Greensleeves."

"Suddenly, Myron's clock, careening madly through the crowd, made a fast left and ran right into Mary, tangling itself into her new shawl, causing a sudden explosion of glittery static. Startled, Mary leaped like lizard, right into my arms.

I somehow managed to catch both her and the smoking shawl without dropping them both."

"So this is how our three paths crossed, with Mary shouting 'HELP!' and me almost toppling over from the impact. All you could do, Myron, was stand there with a sheepish grin. It seems you two have more *oomph* than *oomph-handling*," Bob added with a hidden smile.

Myron blinked. "Bob, are you *speaking* machine?"

Bob chuckled, a low thunder rolling from his chest. "I do, and I *listen* to them. Machines don't shout when you know their bumbling ways."

"And just like that, the three of us knew that together, we could do what none could do alone. Myron with his crackling ideas, Mary with her weaving of warmth and wonder, and me with my grounding, steadying hands. We could build something no one had ever seen before. Something that would not just spark and sizzle, but *sing*."

Bob continued, "Something that could bring light, heat, and music to every dusty village and foggy hollow. Something, perhaps, that could make the world itself become a little brighter." And so, under the crooked beams of Bob's barn — a place that smelled of iron and old dreams — the first plans for the Great Friendly Dynamo were born. A machine not just of parts and power, but of joy, stitched together by sparks, by threads, and by heart.

Breaker Bill entered, adjusted his overalls, and brushed a bit of dust off his mustache, before he said...

"Ah, the next day we went to the Breeze City Power Pow — Pow," and with a spark in his eye, he continued "Now that's a story charged with potential."

He leaned in conspiratorially, as if afraid someone might be listening

"Every year," Bill continued, despite the noisy crowd around them, "the brightest sparks from every corner of the grid gather just outside Turbine Ridge. You've got solar sizzlers, wind wranglers, hydro hustlers, and even a few retired coal cronies trying to stay relevant with brochures. We talk transmission, tackle tariffs, trade tales of power surges and the great blackout of '03. There's hot cocoa made with resistance heating, and the keynote speech is always delivered by Dame Electra Tesla herself — powered solely by applause and ambient Wi-Fi."

"And this year, ***Watt's the Profit?*** was the Pow-Pow theme."

As a door opened, a warm gust rustled the paper on a clipboard in Andrea Amp's hand. She stepped up to the bar, as if she was getting out of an electric Jeep and onto a grassy knoll.

Andrea muttered, "Picture this, thirty-two turbines," squinting at her load curve projections, "producing an average of 3.1 megawatts on a good day — but only when the wind isn't off having an existential crisis." Bill commented as if he were standing on a transformer box. He adjusted his sun visor. "Back in my day," he croaked, "we didn't wait for the wind to flirt with productivity — we made do with hamster wheels and a very motivated woodchuck named Wally."

Bob, wearing his signature "STEP UP or STEP DOWN" utility vest, added solemnly, "The output's decent," he said, pointing at the spinning giants visible through the windows, "but it's not about what you generate — it's about what you connect. Right now, this whole farm is as useful to the grid as a flashlight in a blackout *with no switch."*

Andrea tapped her clipboard. "There's no high-voltage line nearby. We'd need a dedicated step-up substation and a feeder to the main distribution node in Giga Grove."

Bob sighed. "Yea, permits, paperwork, and possibly bribing a very lazy groundhog named Carl Jr. to vacate the conduit path."

Just then, Mary chimed in,

"Ya know, where we were standing among the turbines, I said we could power Breeze City's entire smoothie scene with just three of these — *if* we routed them through one of Bob's step -downs and looped the residual through my nostalgia filter. It adds warmth."

Mary Louise ordered another *Lumenade* cocktail and rolled her eyes. "We're not powering memories, Bob. We're trying to calculate ROI."

Bob, ever the peacekeeper, raised his clipboard like a conductor's baton. "Let's run the numbers, file the easements, and if we need a new substation..." He smiled. "This guy named Breaker Bill owes me a favor and still has access to surplus circuit panels."

And with that, the trio got to work — one equation, one transformer schematic, and one wind-fueled smoothie at a time.

He thumped the ground with his boot. "This year, they're unveiling the Wind Whisper 9000. It can hear a breeze from two counties over and start spinning before the first gust even hits. Now *that's* predictive!"

Myron nodded, impressed. "Do they still have the joule-jumping competition?"

Breaker Bill just winked. "That's the Pow-Pow spirit."

Wendy entered and joined the discussion, "Nonsense,,, here's how it went, were were flying high above the sweeping hills of Breezington, Montana in the electric sky-skipper *Watt's Up One*, when Andrea pointed to Turbine #42, as her eyes scanned the horizon."

Wendy Watt, flying as Pilot, said "I leaned out of the cockpit window, wind whipping through my long brunette hair, shouting, I can almost taste the turbines! Below, the wind farm shimmered like a field of metallic daisies doing synchronized yoga with each swoosh of the blades."

Wendy Watt relaxed at the bar, as she usually did at this time. She was born on the whirling breath of a West Texas wind farm. She is a second generation energy entity — the electrifying daughter of the legendary power team crew Mary Louise Watt, and who helped civilize the raw chaos of early electricity, in addition to being an flight instructor certified in *Electric Aviation Specializing in Vertical Take-off.*

Unlike her mother's era, which wrangled brute-force coal combustion and copper cabling, Wendy was forged from renewables, a photon-turned electron, spun into existence in a solar inverter and given form through wind-powered coding. She's sleek, efficient, with zero emissions but maximum sass.

Wendy, with her fluorescent glow was armed with her *Smart Device Compliance Wand* (which was just a remote with ten thousand buttons, eight of which worked). The smart-phones huddled together, blinking sheepishly. "Sorry," murmured an iPad. "We just wanted to sync one more podcast."

"Yeah," added a smartwatch, "I was updating my firmware. And checking my heart rate. And checking someone else's heart rate too."

Wendy frowned. "Listen, y'all. You're not evil, just needy. But you can't keep sipping juice 24/7. We need a truce. Now everyone, listen to me."

Her talents include whispering to smart grids during peak loads, de-escalating turf wars between solar panels and battery backups, and managing egos of overconfident smart home devices (for which she has a doctorate in AI psychology). This gives her credibility and respect among her peers.

Wendy speaks with the authority of her mother's legacy but runs on a *lighter, greener circuit.*

Where Mary Louise tamed lightning, Wendy dances with the breeze and basks in radiant potential.

She's the ambassador of *The Sustainable Spark* = equally fluent in volts and vibes.

Wendy's resume is broad. She has successfully balanced the modern grid, fought phantom loads, and ensured that no watt, windblown or sunbaked, went to waste.

She said "You all need to hear more about Bob Transformer, Eddie, please tell them more." (Could Bob Transformer possibly be her father?)

So Eddie continued, " **Bob Transformer's** unique skill is turning high-voltage egos into grounded humility. Before electricity can dance politely into your living room lamp or charge your toothbrush with dignified restraint, it has to pass through Bob Transformer - the unsung hero of electrical diplomacy."

"Without Bob, your hairdryer would be a Arc-Welder."

Arc-Welder—Voltage

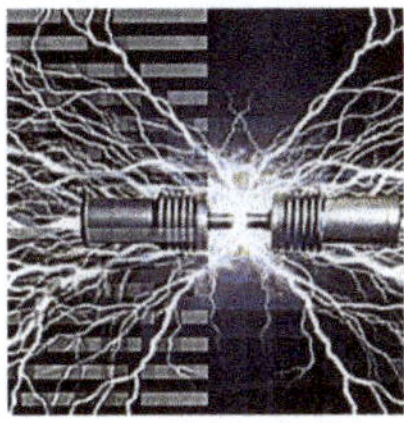

Bob was born in the swirling magnetic fields of Three Mile Island. He had a unique gift from birth, doing cartwheels to step up and step down, both voltage and social situations.

Power Grid Humor on the walls in the Substation Saloon

Bob once stated, "Need a million volts to cross the desert? I'm your guy. Need that same juice gentled down for your nightlight? Still me."

He's not flashy like Eddie Current or as dramatic as Mary Louise Watt — but when the grid starts humming out of tune, Bob's there to balance the buzz and fine tune the tune.

He's the kind of guy who shows up with a clipboard, and a mug of cold dielectric fluid, and says things like: "Let's drop this voltage before someone gets zapped."

Bob is the only one of the Grid Crew to be classified as both "hot" and "neutral" depending on context and he keeps a signed photo of Nikola Tesla in his primary coil.

He is one of the Founding Electrons of The Substation Salon where misbehaving frequencies are sent to cool down.

Hair dryer—Voltage

Just then, **Charlie Capacitor** came out of a back room. Known for his ability to give a Heart of Polarity to the Resistance Collector, he also was good at holding grudges and voltage, then letting go *dramatically.*

Around the Substation Saloon he was the quietest of all the Grid Crew and the *Steady Hand of Power.* While Eddie Current zipped and zapped with high-speed excitement, Charlie was a calming, grounding force. Charlie was a calming deep breath in the relationship of all the Power Team players.

Eddie would come charging into a project, throwing electric energy everywhere like a kid tossing candy at a parade. (You see, Eddie loved *moving* electricity, darting through wires, leaping from coil to coil like a spark on a sugar rush.)

Chalie had a different kind of magic. He could *hold* electricity, give it a hug, save it, then release it just when needed.

Charlie always carried his trusty *Capacitron* , a round, shiny device that looked a bit like a lunchbox crossed with a jukebox. Whenever Eddie needed an extra burst of power (say, to launch the Great Friendly Dynamo into a double-loop spin), Charlie was right there, tapping his Capacitron and sending a pulse of energy back into the system.

Charlie rocked a classic cylindrical look with stylish blue OshKosh overalls, proudly showing off his polarity. One lead on his lip, the other in his right hand, he claimed "I'm emotionally charged. I'm the kind of guy who's calm under pressure, unless you reverse my polarity, in which case *kaboom*."

Quietly, Charlie commented. "I heard that you just got an invitation to play a round of golf at Margo-Logo Golf Club. *Can I be your caddie?"*

"Is that a *Golf Invitation* from our Chief Executive?" **Breaker Bill** asked without looking up, already wiping down a mug that looked like it had seen one too many rownouts.

Eddie dropped onto his usual stool, the one near the jukebox that only played synth-wave and transformer 60 cycle murmurs.

"I played three sets of high-voltage tennis against Andrea Amp. *I was just ionizing.*"

Breaker Bill nodded. "Anything else?"

"I'm tired. Had a stressful day. Before tennis, I redesigned the LED roadway lighting for the St. Armand's Bridge in Sarasota. You know the one. The beautiful arch with terrible sodium colored lighting."

"Is the SRQ City spec is still stuck on that sodium vapor?" asked Bill. Eddie exhaled a low, 30 cycle buzz sigh. "Worse! They said LEDs don't feel authentic. One guy told me humans bond with amber-colored gloom. Another handed me a brochure from 1993."

Then Bill asked "While golfing, are you going to ask for a promotion? Maybe a cabinet position like Secretary of Hi-Voltage Transmission?"

Charlie jumped in with "We'll put a bronze plaque with your name on it on your stool so you'll always be remembered by your friends at the Saloon."

Charlie continued, “Do you know what you need after all this, Eddie? Either a walk through a solar farm or a relaxing game of golf. How about it? A 400 panel lap or 18 holes?”

Eddie groaned. ”Charlie, the last time I played a round of golf, electric meters were spinning out of control. They couldn't even count that high.”

”All the more reason to get recharged!”

At that moment, the bar lights began to dim and our crew saw a young electron with a harmonica in one hand and a tall bar stool in the other setting up to play a few lively sets.

“Now this is the kind of relaxation I like,” Eddie thought, a smile replacing the frown on his face.

Meanwhile, from across the bar, Breaker Bill intoned in his characteristically deep, rolling voice, “Golf was invented by the same folks who believe great music is best played with bagpipes.”

The small talk and chatter in the saloon covered many topics that evening, from lawn darts to horseshoes to hop scotch. Eventually, Social Media came up and the role it plays in creating todays news and how this affects human lives as a source for all current news and information.

Breaker Bill popped his head in and started talking about the influence of Social Media channels and the news they were spreading. With that, the lively conversation at the bar suddenly fell silent.

After a long pause, Breaker Bill broke the silence saying "Yeah, there are youngsters today who take everything they see on the grid for real."

In truth," Eddie continued, "youngsters and adult humans alike scroll through endless posts, memes and mayhem for hours on end. But they need to beware! Among the pearls of wisdom and the funny flicks lurk nasty critters known as *Fake Newsies.* They sound convincing but spread misinformation. Some tales are taller than a giraffe on stilts! If only we had the power to cast the ancient spell of *Critical Thinking +2* to break their algorithmic addiction!

Fortunately, our brave young electrons, driven only by surge and armed with pure power, grow ever more enchanted by the possibilities the future holds for us and for humanity alike. Go kids!

Power Team and Social Media

On one very sunny afternoon, the young electrons were feeling unusually photogenic. It was their first annual outing to the Sunnyvale Solar Farm since the Covid lockdown. They liked the farm because it was a sprawling field of glittering panels with a few wind turbines along the edges and endless possibilities.

As they basked in the photon frenzy, the Social Media Set arrived with full algorithmic flair. TikTok twirled in a whirlwind dance routine, Instagram struck twelve poses on a sunbeam while adjusting her filter, and Twitter hovered above, squawking hashtags from a lightning drone.

LinkedIn arrived late but professionally dressed, passing out pixelated résumés, while Facebook set up a nostalgic photo booth full of old selfies.

YouTube brought a drone crew and started livestreaming the whole event. They promised to "amplify the vibes," "energize engagement," and, according to a suspiciously enthusiastic Pinterest board —"solarbrate the moment."

Just as the selfie drones began forming a hashtag in the sky (#SolarVibes), a sleek buzz filled the air — and in soared Andrea Amp, gliding on her glitter charged hoverboard. Her suit shimmered with conductive flair, as her clipboard pulsed with circuit-synced insights.

She landed gracefully between two Instagram Reels and raised an eyebrow at LinkedIn, who was mid-monologue about "leveraging photovoltaic synergy for cross-platform brand cohesion." Andrea muttered, "LinkedIn always sounds like it's applying for a job it already has." Then, flashing a smile that could outshine a solar inverter, she added, "But bless its bandwidth — it's the only one here who brought a mission statement."

The electrons gave a collective sizzle of agreement as Andrea adjusted her flux shades and began cataloguing wattage-to-engagement ratios, proving once again that no solar shindig is truly powered up until Andrea Amp plugs in.

Breaker Bill and the Social Media Time Warp

Breaker Bill had heard many stories of the construction and benefits of Sunnyvale Solar Farm, and should have felt welcome, yet our Substation Saloon bartender felt *disconnected.*

Eddie recalled "Long ago, before the rise of all the different Social Media channels, before likes, shares, and un-skippable ads that appear, Bill, I remember you saying 'back in my day I got my news from *four trusty sources.*'"

* "**Newspapers**, those ink-stained truth grenades which were dropped on countless porches at 5 AM sharp.

* "**Radio**, where disembodied voices told us the weather and what Elvis would do.

* "**Television**, where just a handful of trusted anchormen delivered global affairs directly into my shag-carpeted living room precisely at six every night.

* "And of course, **Neighbors,** my real ones, the kind who leaned over my fence to talk politics, crops, and best of all, juicy gossip."

Twitter/X followed: *"Wired and Weird in 280 Volts" or "The Chaotic Oracle Where Sanity Goes on Vacation."*

Eddie took to X (formerly Twitter) with chaotic brilliance: "If I get unplugged one more time while transmitting joy, I *SWEAR* I will become a lightning bolt and haunt your vacuum." @RealEddieCurrent

His tweets became electric haikus**

Circuit to socket
Resistance is overrated
Plug me in, baby

He was blocked by four home insulation outfits but followed by NASA.

Then, it was **YouTube's** turn: *"Welcome to Current Events with Eddie"* Eddie launched a video series.

First episode: *"In Reality, Does a Neutral Wire Exist?"*

Second episode: *"Do Toasters Dream of Electric Crumbs?"*

Third episode: *"Microwaves: Science or Sorcery?"*

His subscriber base grew to 43,000 before he accidentally live-streamed himself doing cable yoga in a surge protector.

Still, the comments were mostly positive.
One viewer wrote: *"I've never been so entertained by circuitry. My bread is toasted AND my mind is blown."*

Eddie finally found **LinkedIn:** *The Office Party That Never Ends*

He logged into LinkedIn and was immediately bombarded with:

"So excited to start my new job as Chief Synergy Officer at Smorb Inc.!

"Humbled to be awarded the Golden Stapler for Outstanding Spreadsheet Vibes.

"If your CEO doesn't cry at meetings, are you really innovating?"

Eddie didn't understand a word. But somehow, he endorsed a goat for "Ethical Leadership."

By week's end, Eddie was:

* Verified on all platforms,

* Invited to do a series on *The Ethics Of Zapping* for TED-Talks,

* And offered a deal to host "Dancing with the Diodes."

Yet through it all, and in spite of many Elect*ric Phenomena*, Eddie stayed grounded.

*** A Japanese poem traditionally evoking images of the natural world.*

Later, Eddie Current Goes Viral
Fast-Forward to the Digital Awakening...

"Everyone's talking about going viral," Eddie muttered, sipping his electrolytic Voltage-aid.
"I *am* electricity! If anyone should be going viral, it should be *me!"*

So, with a determined spark in his eye and a faint buzzing in his boots, Eddie plugged in his Smart Phone and logged on to all the networks he could find, starting with —

Facebook: *"The Circuit Board Reunion Club"*
Eddie's first post was a heartfelt memory from his days wiring up a retirement center's solar array.

"Feeling nostalgic — Shoutout to the 2012 Golden Watt Club! It warmed my diodes."

Five comments. Three likes. One reader (possibly a wall outlet in Kansas) tagged him in a photo of a junction box.

Eddie replied, "Stay grounded, friends!" and moved on.

Next came **Instagram:** *"Shockstagram Influencer Mode: Activated."* He posed dramatically next to a Tesla coil, captioning it: *"Feeling charged. New circuit, who dis?"*

#WattAFit
#ElectroAesthetic
#SparkingJoy

Within minutes, he was followed by 12 electricians, 4 confused circuit DJs, and one motivational lightbulb named Glowlene.
He quickly developed a niche: aesthetic wiring shots, fusion selfies, and grid-positive cool quotes like:

"Current mood: alternating."

Then **TikTok** came up on his screen — the neon rave of the digital realm. *Within seconds he was hooked.*

One guy taught Eddie how to dance with a mop. Another whispered life advice while making Chicken Chili. He was learning *nothing*, but *everything* felt urgent. Eddie blinked. It was 3 AM and he had a sudden urge to buy frog-shaped soap. TikTok had won.

Eddie saw:
A guy whispering financial advice while slow-dancing in a Walmart.

A grandma making lasagna while ranting about moon phases.

A teenager teaching the "Quantum News Shuffle" while lip-syncing to sea shanties.

And, was offered a deal to host "Dancing with the Diodes."

Suddenly, Eddie saw a clip of *himself*, falling asleep in his recliner. It had 6.2 million views... **WOW, is this what fame looks like?**

He blinked. The app asked: "You've been watching for 4 hours. Are you okay?"

"I'm Great!" he replied, undeterred as he moved on.

Electric Phenomena: Power Team and Grid

The following week, the electrons called a formal association meeting at the Substation Saloon. It was standing room only at the copper bar and the large hall buzzed with excitement.

One electron whispered to his neighbor, “Isn’t today’s topic Electric Phenomena and how They Affect Us? Boy, this should be interesting!”

That’s when the double doors swung open and in strutted Zaparella, the crackling queen of conductivity, trailing arcs of static like a sequined thunderstorm, as she approached the bar. She tossed a lightning bolt into the tip jar with flair and declared in a commanding tone, “Let’s get this circuit in session!”

Behind her strutted five Darkness Suckers — Incandescent Inez, Fluorescent Flo, Sodium Sal, and the ever-mysterious LED twins, Lumi and Diode.

Each had a signature style of banishing gloom by slurping espresso-like darkness straight from the tap. (No cream, no sugar please.) They took their usual booth under the flickering exit sign, nodding with gravitas. “Let’s review the darkness extraction reports,” said Zaparella, dramatically unrolling a chart that glowed faintly with residual gloom.

The meeting was lit — literally — with the electrons, now properly charged and slightly overcharged, prepared for another shocking session of great importance to them all.

DARKNESS SUCKERS

ON

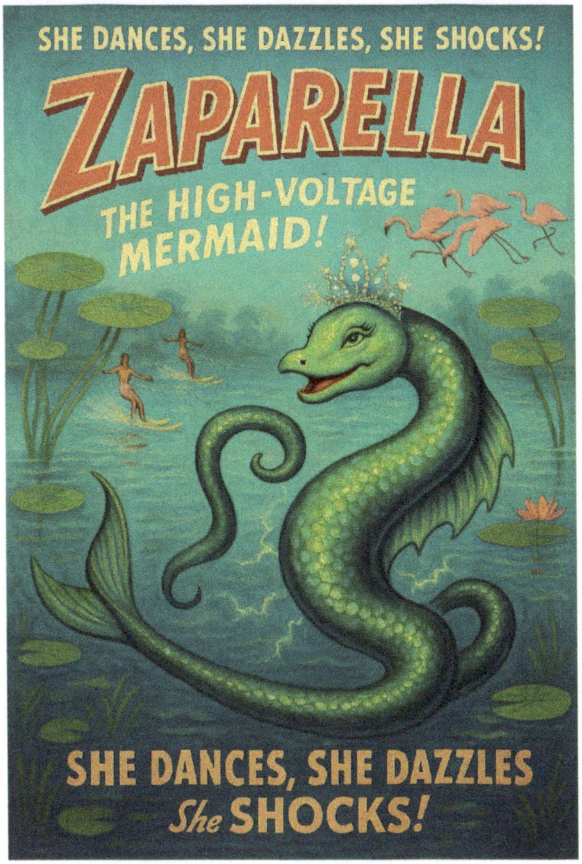
SHE DANCES, SHE DAZZLES, SHE SHOCKS!
ZAPARELLA
THE HIGH-VOLTAGE MERMAID!
SHE DANCES, SHE DAZZLES
She SHOCKS!

The first speaker (born in a thunderstorm) stood up, cleared his throat and began the meeting in a serous tone.

"It's an established scientific fact that ***Darkness Suckers*** gather darkness from a room, convert it into radiant energy, and then send it through transmission lines like a fine cappuccino in a copper straw. Once back at the power plant, the concentrated darkness is responsibly buried beneath wind turbines or solar panels, from where it can't break out. The first warning sign of trouble is when a bulb develops black ends. If it does, call your local Department of Luminance."

"It means your darkness sucker is full and may begin leaking shadows into the corner of your laundry room. If you hear faint whispers or repeatedly lose socks, darkness may be seeping into your home."

Since there were no questions or comments, he continued.

"**Darkness Suckers** are sneaky little things, quirky, a little absurd, slightly domesticated and tinged with mystery, like reverse sparks that slither out of the walls when no one's looking. You know that tiny fizz you hear when the toaster clicks off? That's a darkness sucker clocking in. They don't chase away light; they *drink* it — casually, unapologetically, as if it were lukewarm coffee left unattended.

"Long ago, when the world was young and shadows ran wild, a noble league emerged to battle this inky menace, The *Darkness Suckers*. Sworn to serve homes, warehouses, parking lots, and that weird hallway behind the utility closet, these tireless champions had one mission — to suck the darkness out of alleys and hidden spaces, to keep the world well lit and moderately cheerful.

"They buzz in the circuitry and stretch themselves through lamp cords, sipping d*arkness* through invisible straws. One moment, your room is filled with shadows, the next, a darkness sucker has inhaled, and those shadows grow bright, room edges get sharp.

"They're not evil. They're just efficient light recyclers, collectors of cozy gloom, giving shadows a place to stretch their legs.

"The First Suckers were of The Incandescent Order.

"Sir Edison Incandescent, the CEO sucker, was a brave glass bulb filled with a glowing filament of determination and a dash of noble arrogance. He'd heat up until he practically sweat light.

"But make no mistake — his mission wasn't to shine, oh no. His job was to inhale darkness with a hiss, a gurgle, and add an occasional pop. When his ends turned black, it meant he was stuffed full of Grade A, locally sourced *darkness*.

"*'Retire me, good sir,"* he'd wheeze. *"I am full.'"*

The speaker continued, **"Incandescent Darkness Suckers** were the first noble pioneers in the modern day eternal battle against gloom. These brave bulbs were among the first to take up arms (filaments, really) against the creeping shadows of unlit rooms and dusky basements. Their mission was simple: suck the darkness out of a space and convert it, miraculously, into a warm, golden glow. Guilds were formed based on their DNA.

"**The Incandescent Guild**—Each bulb contains a tiny tungsten filament, a fragile little coil that glows heroically when electricity courses through it. This filament burns with passion (it also sucks cold) in its quest to banish the dim and dark murkiness.

"The Fluorescent Guild — Next came the Fluorescents — tall, glassy tubes glowed like spooky ghosts and hummed like distant bees. Fluorescents didn't just suck darkness — they slurped it in stylish, flickery pulses. They worked best in hospitals, garages, and unsettling office bathrooms. But once their tips turned black? 'Oh dear,' they'd whisper. 'We're at capacity. Please dispose of us improperly so we can haunt landfills forever.'

"The Merc Vaper Bunch — Mercury Vapor lights were grizzled veterans of industrial areas. They had no time for nonsense. Wearing blue-green auras
like battle scars, these suckers roamed the perimeters of warehouses and factories, ready to pull darkness out of the air like seasoned ghostbusters.

"**The Sodium Knights,** orangey and mysterious, wandered foggy roads and parking lots, muttering poetry and sucking darkness with a golden glow that made everything look like it came from a 1970s detective show.

"**The LED Enlightened.** Enter the *LEDs*, sleek, smart, and strangely chipper. These weren't just suckers — they were darkness alchemists. Sadly, they were often only judged by their efficiency. Along came the cold, calculating LEDs. The smug suckers of today are pushing our heroic friends to the dusty shelves of nostalgia.

"And so, the fight continues. In every lamp, ceiling fixture, and dimly lit stairwell, the Darkness Suckers keep watch — protecting humanity from the creeping tendrils of gloom, one watt at a time."

Zaparella grabbed her poster and said "enough talk about Darkness Suckers, what's next?"

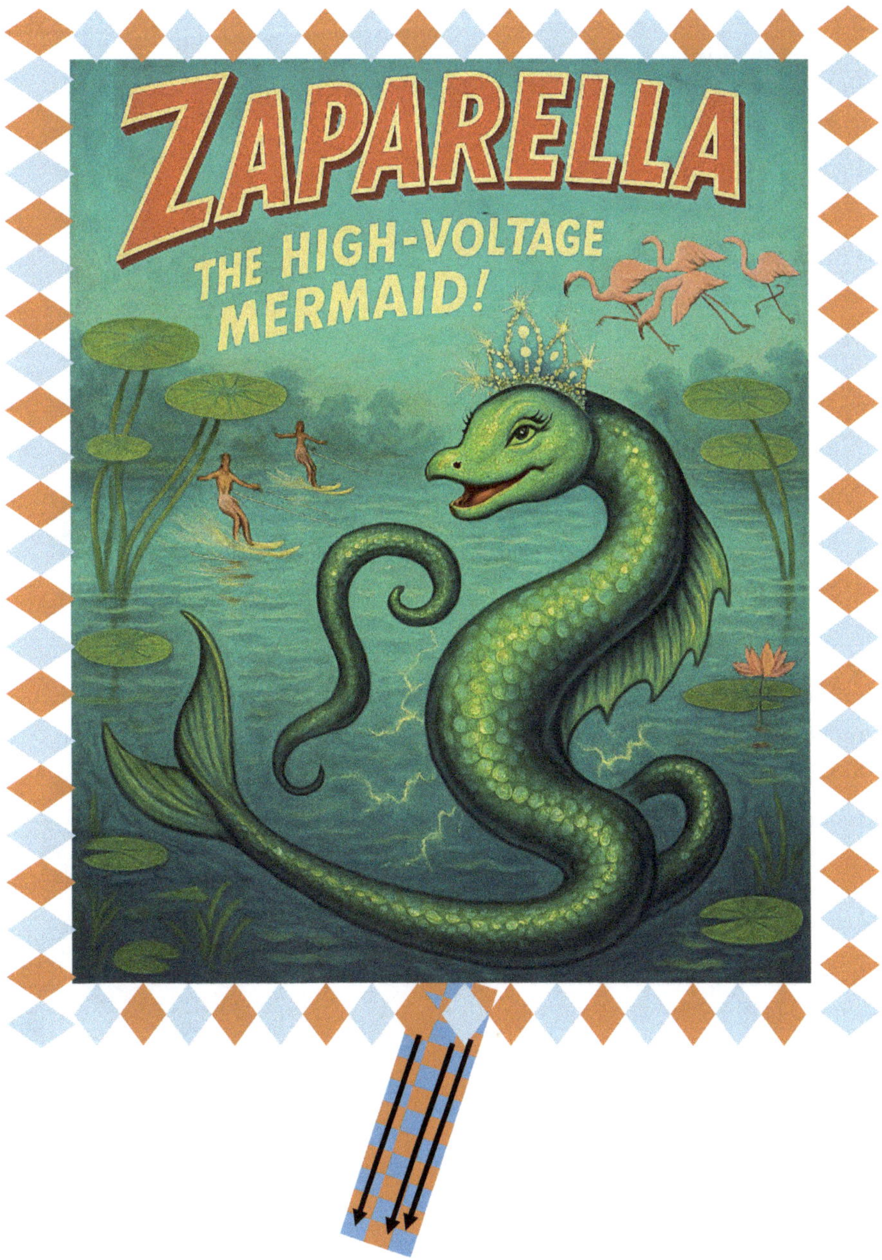

The speaker continued, "Why don't we let Eddie tell his story about Zaparella."

The bar filled with loud applause as Eddie came up to the podium with a huge smile, Zaparella slithering behind him. "Sure. I'd love to."

He began, "Here's how I met my **Electric Eel.** One bright morning charged by solar, of course, I decided something was missing in my life.

"Was it a fresh pair of insulated gloves?
A new smart meter?
Or a stronger Wi-Fi signal?"

"No. It was... companionship," he declared, dramatically unplugging a toaster for emphasis.

"I needed a pet," he stated.

"I tried the traditional route:
A *goldfish?* Too analog.
A *cat?* Too smug.
A *dog?* Too grounded," Eddie sighed.

"That's when, on a visit to the sun glittered,
lily-pad-strewn waters of Cypress Gardens, Florida,
once the twirling jewel of synchronized swimming films
a place where water ballet met jazz bands and every
splash sparkled with musical ambition,
I met her! Zaparella, *an Electric Eel*!"

With that, Zaparella smiled, teeth flashing. She had entered the competitive swimming circuit. She wasn't just any contestant. She was performing solo and fully charged. Eddie's eyebrows rose, metaphorically...

"I knew instantly, in a nanosecond, that this was the perfect companion for me," said Eddie.

"But, how could I get her?"

"So I asked a nearby fisherman. He looked looked nervous. 'She discharges when startled. Or bored. Or happy. Or when jazz music plays.'"

"I grinned. 'Perfect. We'll get along swimmingly.'

"And so, Zaparella was adopted asd moved in.

"Life with an Electric Eel was magical at first. She helped me start the coffee maker. She powered my blender. Once, during a blackout, she recharged my entire backup battery bank just by swimming in tight circles while whistling the theme from *Flashdance*."

"But she has quirks," Eddie continued.

"She loves to hide in the dish-washer. She re-wired the door-bell to play "ThunderStruck," and she has opin-ions about my extension cords, particularly the green ones. "Worst of all, she *vaporized* my smart TV!"

"Nevertheless, a shocking bond formed between us. I couldn't deny it, we were a perfect match!

"Every morning I would toss her a frozen voltage fish stick and she would flick her tail in gratitude, lighting up the kitchen like a Las Vagas Casino sign. I even gave her a tiny red cape to wear.

"We traveled together. We toured substations and once did a TED Talk called *Bioluminescent Bonds: Love at First Amp."*

"From that first day on, I never felt alone again, because sometimes the best pet isn't furry, doesn't fetch sticks and hates to be scratched behind her ear. "Because of Zaparella, young girls now play with eel-shaped dolls and boys have metallic green eel transformers."

And so, beneath the tall Cypress trees, in the sizzling Florida sun, the worlds first pet electric eel has made her way into our hearts, made headlines and entered the halls of hydroelectric history.

Unfortunately, the world eventually realized that electric eels can be slithery, slightly dangerous and that they glow a spooky green under black light.

All electrons gave her wide berth whenever she was in the area."

A young electron jumped up and asked the speaker "Are **Beacon Lighthouses** a phenomenon?"

The speaker then turned on a projector with a video recently recorded from *Facebook*. "Long ago he began, the high seas were ruled by pirates — ruffians with tricorn hats, poor impulse control, and a deep mistrust of GPS. By day they plundered and pillaged with reckless glee, but by night, their galleons often crashed into the dreaded rocky shoals of Cross-Arm Bay which was named after a particularly bad transmission pole.

"That all changed when it was discovered that the Power Grid offered not treasure but Beacon Lighthouses, powered by pure, steady, sailor-saving electricity. Now, thanks to circuitry wisdom and a few well-placed lightbulbs on sea-splashed lighthouses, even the most directionally challenged buccaneers can steer clear of danger."

"*'Yo ho ho, Yo ho ho,'*" they sang, "'It's a bottle of rum and a *lit* life for me!'"

An Interesting Fact-Check:
Could this be Social Media Fake News?
The Golden Age of Piracy was the period between the 1650's and the 1830's, when maritime piracy was a significant factor in the histories of the North Atlantic, BUT,,, Ben Franklin is credited with flying the kite in 1752 although the first use of electricity at a US lighthouse wasn't until 1884, at the Hells Gate Lighthouse in New York. However, this was discontinued due to the light being too bright. The first "permanently" electrified lighthouse was the Statue of Liberty in 1886.

The narrator electron asked, "Any other questions?"

A voice in the in the back called out, "Some might consider the **Drive-in Theater** an Electric Phenomenon. Back in the 1950's and 60's, it wasn't just a place to watch movies — it was the unofficial headquarters of young love."

An old-timer electron sitting, half-asleep added, "I remember rows of chrome-trimmed cars lined up under the stars, each a private cinema and personal bubble of possibility. We young romantics practiced the ancient ritual of accidentally brushing hands over the popcorn bucket, while monster movies and melodramas flickered on huge screen. Beneath the drone of the FM signal and the soft rattle of speaker boxes, our hearts accelerated faster than a Ford Fairlane leaving a burger stand.

"Whether you came for the film, the romance, or just to impress someone with your parallel parking skills, the drive-in was where sparks flew, especially if Eddie Current was on projection duty that evening."

"Were these **Electric Phenomena?** Let's vote," said Breaker Bill.

"NO, NO,, NO,,, *NO!!!!"* was the overwhelming response!

The Work
The Power Team
Does For Us

The **Power Team of the Grid** isn't just loaded with personality — they also pack an electrifying arsenal of home brewed gadgets, gizmos, and grid-hacking gear.

Leading the charge is Charlie Capacitor's Capacitron — a shiny, round, clicky lunchbox-turned-jukebox that stores emergency energy. Mary Louise Watt created the Ohm-O-Meter, a sleek wrist gadget while Bob Transformer hauls around the Voltage Valise, a briefcase that can step up small talk or step down panic in a blackout. And let's not forget Eddie Current's Circuit Cape.

Together, these tools make life fabulous, and, on Ted-Talk, *The Shocking Truth About Electrified Living,* we will explore several of the everyday wonders that bring us joy and convenience.

Eddie Current said (grinning): "Ladies, gentlemen, smart devices and that one toaster in the third row — welcome! I'm Eddie Current, and I'm here today with the unsung heroes of your daily life: The Power Team of the Grid."

⇒ **The TenderZap™ 400**

⇒ **Power-Hungry Heating Pad**

⇒ **The Electric Chair:**
A Shocking Seat of History

⇒ **The Smart Phone**

⇒ **Enchanting World of Lasers**

⇒ **The Electric Household with Router**

⇒ **The Electric Screwdriver**

⇒ **The Phantom Load**

⇒ **Theater: Live & Electrified**

Tender is the Zap:

The Shocking True Story of Culinary Ambition

Stosh "Sparky" Pendosky (not a Power Grid associate at all, just a tinkerer) loved a dinner of grilled steak and considered himself a kitchen innovator. The air fryer? Old news. Sous vide? Yesterday's steam. No — Sparky had bigger dreams, *higher voltage* and a ribeye with the texture of a medieval saddle.

His first test subject: a 16-ounce sirloin and a slightly reluctant hungry neighbor named Carl, who agreed to supervise only after being promised steak and *absolutely no burns this time*.

"Let me demonstrate, Carl " Sparky said. He clipped the steak to the TenderZap™, flipped the switch... and for exactly 0.7125 seconds, the kitchen was illuminated by a holy light not seen since the toaster rebellion of '08. The steak levitated. The air smelled like ozone and beef betrayal. Carl dove behind the fridge.

When the smoke cleared, the sirloin was... gone!

Carl had not seen anything like this since the toaster rebellion of '08.

"Behold," he announced, unveiling his newest invention: The TenderZap™ 4000, a re-purposed electric fence, a meat hook, and what looked suspiciously like jumper cables from a beat-up, Raw-Meat-Red 2002 Ford Taurus.

"We don't *pound* meat anymore," he declared. "We *persuade* it. Electrcally."

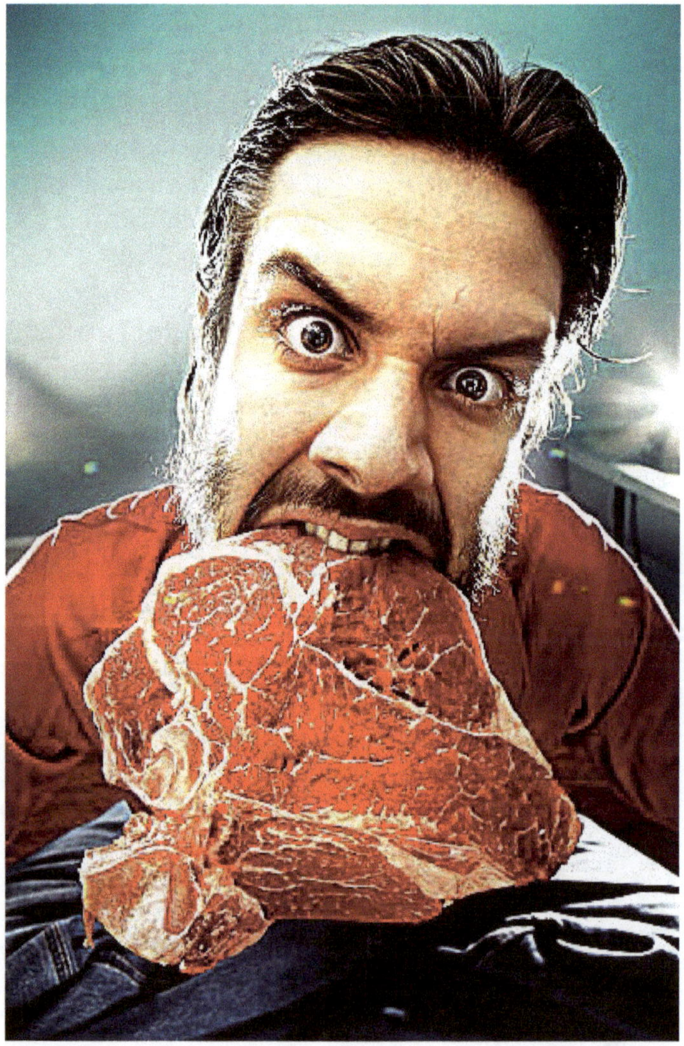

They later found it stuck to the ceiling fan, fully cooked, slightly charred, and eerily humming the opening bars of *Bohemian Rhapsody*.

Sparky took a bite. “Perfect medium-rare. A little *tingly*, but elegant.”

Carl, clutching an extinguisher, muttered, “You built a meat defibrillator.”

“No,” Sparky corrected, wiping his lips.
“I built the future.”

They spent the next hour unplugging everything in the house and writing *very* careful instructions for the **TenderZap™ 4000**, including the now-mandatory warning:

*****Do not use near water, pets, or in the presence of God*****

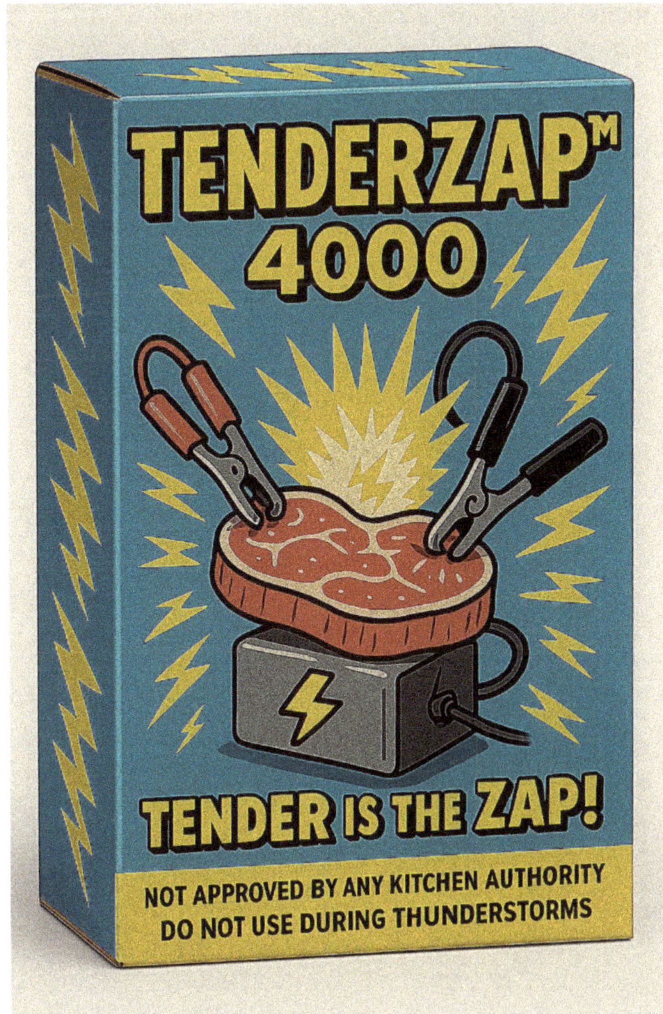

Case of the Power-Hungry Heating Pad
Eddie Current decided he needed a break.

After a grueling few months of transmission line wrangling, tariff tinkering and an unfortunate incident involving a rogue cat in a substation, Eddie's lower back was officially fried. Not electrically, mind you— just the classic "you lifted a 3-phase step-down transformer with bad posture" kind of fried. So, Eddie booked a cruise, a seven night voyage through the Caribbean aboard the ***S.S. Ampere Dreams***, the only ship that promised both unlimited buffet and dedicated lightning rod protection. But he forgot his suntan lotion. He packed light: flip flops, a voltmeter, backup socks, (also forgot his bathing suit but not his tool belt) and most importantly — his trusty heating pad, affectionally named "Toasty."

Day 1: *Boarding Denied*... sort of. As Eddie boarded the ship, a cheery crew member named Doug, dressed in a polo shirt, asked routine questions. "Do you have any firearms, flammable liquids, or high-drain electrical devices?" Eddie hesitated. "Uh... just a heating pad. Old school. Low resistance. Sizzles like a dream."

Doug's smile faded faster than a capacitor discharging through a resistor.

"We're going to have to confiscate that, sir. Captain's orders. Heating pads pull too much current from the ship's grid. Last week someone tried to plug in a fondue fountain and shorted out the shuffleboard scoreboard."

Day 2: *The Great Negotiation...* Eddie wasn't one to give up so easily. He met with Captain Ohmstrom himself — a stoic sea-skipper with a stern brow and a tattoo of a waveform on his bicep. "Captain," Eddie said, placing a kilowatt-hour chart on the captain's desk, "my heating pad draws only 0.4 amps at 120 volts. That's *barely* enough to tickle a toaster."

"Oh, I know your type," the Captain growled. "One heating pad leads to an electric blanket. Next thing I know, we've got passengers plugging in mini fridges and induction cooktops! The ship's grid is delicate. She's designed for karaoke, not kilowatts!"

Day 3: *Underground Circuitry...* Undeterred, Eddie befriended the ship's onboard electrician, Lucille "Loop" Magwire. She guided him through the bowels of the ship to a hidden utility closet which pulsed softly with ambient energy.

"I can wire you up to a temporary circuit," Lucille whispered, slipping him a fuse like it was contraband. "But you didn't get it from me."

Day 4: *The Captain's conference...* Dave and Lucille answered the Captain's order and returned the heating pad to Eddie, as he said "your detailed conversation must have been very convincing, but I'm not convinced." Dave then cut off the cord in an angry state. Eddie didn't mind, he knew he could rig a substitute, so he just smiled and said "thank you."

That night, Eddie powered up "Toasty" with a sigh of relief and promptly fell asleep on his bunk, smelling faintly of aloe and ozone.

Day 5: *Disaster Struck...* Grid Failure at Midnight struck during the midnight limbo contest. Just as the reigning limbo champion reached ankle-height, the ship's lights dimmed and a disco ball tragically stalled mid-spin.

Sirens wailed. The intercom crackled. "We are experiencing an unexpected voltage drop. All systems are being rerouted. Remain calm and do NOT microwave anything."

In the control room, readings showed an unexplained energy sink. All eyes turned toward Eddie's cabin.

Day 6: *Trial by Panel...* Eddie was brought before the ship's "Council of Cruise Conduct." In his defense, he drew schematics in mustard on a napkin and argued that the real culprit was a group of teenagers using hair straighteners in unison.

But the Captain was unmoved, stating: "Eddie, it appears that you have a Flexible Moral Compass."

"You violated the One-Watt Rule," he said gravely. "You'll be sentenced to community karaoke until your voltage debt is repaid."

Day 7: *Redemption and Rum Punch...* Eddie sang *"Electric Avenue"* with surprising gusto. He even fixed the karaoke machine mid-verse when it tried to loop endlessly on a B-flat.

His punishment was commuted to "light DJ duty," and the Captain eventually relented and let him plug Toasty in again — but only at night, and only if he agreed to insulate it with a pool noodle.

Post-Cruise Reflections:
Back on dry land, Eddie's back was still sore, but his spirits were charged. He had survived the Caribbean, made new friends, and even got a tan (though oddly shaped — partially due to the coiled cord of his heating pad).

He made a vow: Next vacation, he'd stick to somewhere simpler. Maybe just a nice, relaxing tour of a geothermal plant or an off-shore wind farm. He then had another thought, maybe the next vacation would be to a battery storage facility, like one of those that converted water to ice then let it melt, providing electric current in the process, slowly, quietly, releasing it when needed without anybody watching. He thought, "I need to call *Danney's Discount Travel & Cruise* to find out where there might be a good deal for next year."

Electric Chair: A Shocking Seat of History

I knew you would ask about the electric chair — a contraption that sounds like it belongs in a mad scientist's parlor or perhaps as the center-piece of a particularly avant-garde furniture collection. But this isn't your average armchair; it's a hot seat with a truly electrifying past.

Imagine a chair so electrifying that it could make your hair stand on end — permanently.

In the late 19th century, Dr. Alfred P. Southwick, a dentist from Buffalo NY, witnessed a man die instantly after touching a live electric generator.

He envisioned a more "humane" method of execution than hanging, which led him to the creation of the electric chair.

Southwick, accustomed to dental chairs, naturally designed this new device in a similar fashion.

The first execution by electric chair took place in 1890, marking a new chapter in capital punishment. While intended as a more civilized alternative, the method sparked debates about ethics and humanity.

So, while the electric chair might not be the kind of seat you'd want in your living room, its history is a testament to the lengths society has gone in its quest for progress — however bizarre those methods may be.

The first person to experience this electrifying invention was William Kemmler in 1890. His execution was intended to be swift and painless, but it turned into a gruesome spectacle, with reports of his body catching fire. Despite its grim debut, the electric chair became a symbol of modern execution methods, earning nicknames like "Old Sparky."

Over time, its use declined in favor of methods deemed more humane, but its legacy remains a testament to humanity's electrifying attempts at progress.

So, while the electric chair might not be the kind of seat you'd want in your living room, its history is a testament to the lengths society has gone in its quest for progress — however shocking the methods may be.

The Smart Phone: A Life Support Machine

Advancing to a more pleasant time, we leap into the 21st century.

We used to roam the earth freely, guided only by stars, paper maps, and the occasional gut instinct — now we panic if our *Smart Phone* drops below 20%, like it's a secret signal that only we know the meaning of.

These magical rectangles have become our calendars, flashlights, cameras, personal chefs, therapists, GPS overlords, banks, and social approval dispensers.

Forget water — if you really want to see someone spiral, just whisper *"No signal."*

We consult our *Smart Phone* even before making eye contact with another human, and before deciding what to eat, and sometimes even before remembering our own names. We walk into a crowded room and the first thing we grab, you know it, is our *Smart Phone*.

Without them, we're just glorified apes
with credit cards and no idea where we parked.

Power Grid Humor
on the walls in the
Substation Saloon

Laser Larry: The Beam with a Dream

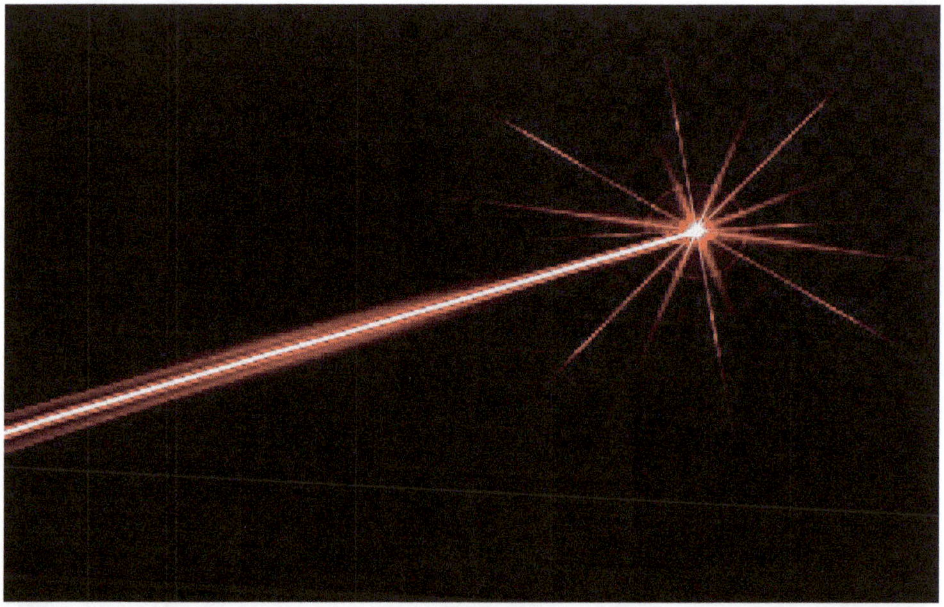

"This is madness," grumbled ***Laser Larry,*** the electric laser tape measure, as he squinted through his precision lens.

"The angles are an insult to geometry." And through the magic of Bluetooth, he transmitted data to a nearby iPad. Apart, they were just tools. Together, they were an amazing team.

A team that brought order to chaos, elegance to entropy... and *just the right amount of sass to complex problems.*

Imagine electricity as a master illusionist, conjuring beams of light that can precisely measure distances, slice through steel and delicately sculpt the human eye. Welcome to the mesmerizing realm of lasers!

And this is just the beginning...

Lasers: Electricity's Grand Performance

Industrial Lasers: The Mighty Magicians

In the industrial arena, lasers are the strongmen of the show. Fueled by high-voltage electricity, these beams can cut through thick metals with precision and ease. It is electricity wielding a light saber, transforming raw materials into intricate designs.

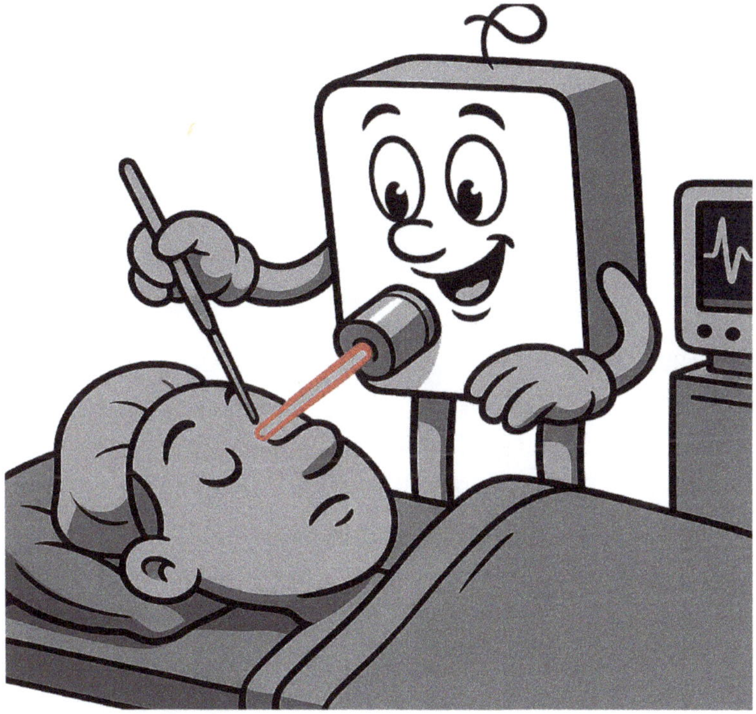

Medical Lasers: The Gentle Artisans
On the flip side, lasers also perform delicate tasks in the medical field. In procedures like LASIK, lasers reshape the cornea to correct vision, offering patients a clearer view of the world. It's a testament to electricity's finesse, turning light into a tool of healing.

Electricity: The Versatile Virtuoso
Whether it's powering a laser to cut through titanium or to fine-tune the human eye, electricity is the unsung hero behind the curtain. Its ability to adapt and perform across such a broad spectrum showcases its unparalleled versatility.

So, the next time you witness a laser in action, remember: adjust the power setting to suit the task at hand. Whether it's vaporizing a bulldozer or performing delicate eye surgery, it's not just science — it's magic powered by electricity!

Ralph the Router

It was mid-week. A slow news day. Peace rang out across the house. The grid pulsed gently, no longer strained.

Reflecting on past achievements, **Ralph the Router** had once been the heart of the home, the gatekeeper of the internet, the maestro of megabits. Now, he was relegated to the shadows, replaced by sleek mesh networks and cloud-based systems. Today, nobody even gave him a thought.

"Back in my day," *Ralph the Router* grumbled, "I was the only one who knew the Wi-Fi password. Now, even the fridge has its own IP address!"

The other devices paid him no mind. The smart thermostat adjusted temperatures with smug precision. The voice assistants chatted away, responding to every whim of the homeowners.

But *Ralph* had a plan.

Late one night, when the house was silent, *Ralph* initiated Operation Reboot. A surge of data packets coursed through the network, causing lights to flicker and devices to stutter.

The smart TV turned on by itself, blaring info-mercials. The robotic vacuum began a chaotic cleaning spree, knocking over furniture. The smart fridge displayed cryptic messages: "I am cold. So very cold."

The homeowners awoke in confusion, their once-harmonious smart home now a cacophony of technological tantrums.

Realizing that the chaos stemmed from the network, they ventured into the utility room. There, they found the *him*, blinking defiantly.

With a sigh, they dusted him off and gave him a firmware update.

"Welcome back, old friend," they said.

Ralph purred with satisfaction. Order was restored for now. But far, far away — deep in a forgotten corner of the utility room, a dusty, blinking box stirred. I may be old. A bit bitter. But *I'm still* the same router that was installed when the house was built.

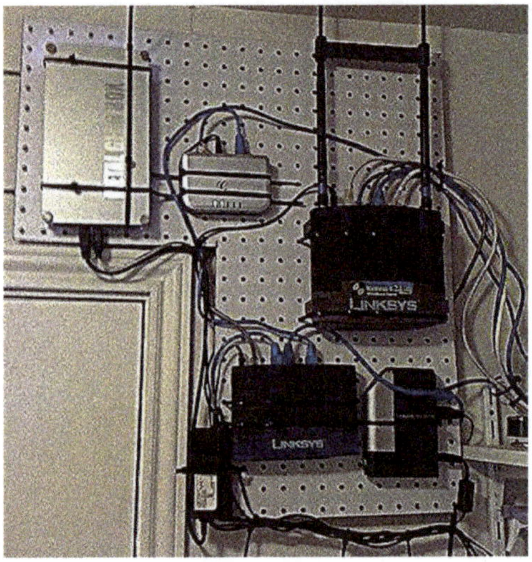

Moral of the story: Never underestimate the power of legacy hardware. They may be old, but they still have a few tricks up their ports.

The Electric Household

The Electric Household is a buzzing ecosystem where every appliance, gadget, and glowing button hums in synchronized service. From the toaster that crisply browns your breakfast with pride, to the Wi-Fi router brooding in the utility room like a forgotten wizard, including the printer for the computers, a security system, music speakers in every room, gadgets to control the lighting, the appliances, even the refrigerator.

Each device plays its part in a symphony of circuits. Guided by the ever-watchful electrons — and occasionally supervised by Eddie Current and friends — this home doesn't just *run* on electricity... it *lives* on it.

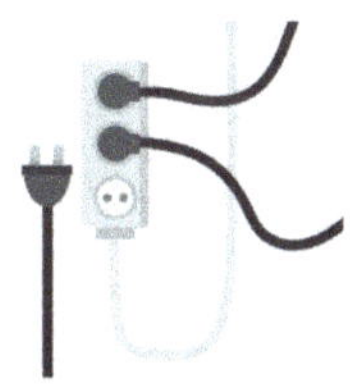

Stanley The Electric Screwdriver

Once upon a plug-in time, in the back corner of a dusty garage, lived a jittery little tool named *Stanley, The Electric Screwdriver*. He wasn't the strongest, nor the most organized (he frequently forgot where he left his own bits), but by golly, he had *enthusiasm*. Other tools scoffed at him. "Leave it to the drill!" barked Big Red.

"You're too twitchy for fine work," mumbled Harry the Hex Wrench. But Stanley knew — *deep down in his rechargeable battery pack* — that he was meant for something great. His motto was ""Yes, you could do this manually... but why?"

Then, one stormy afternoon, a crisis rocked the house: THE CABINET HAD GONE WOBBLY!

Utensils rattled, plates clinked, and the entire silverware drawer was leaning like it had just returned from a blender party. The home owners gasped. The dog judged silently.

The older tools had retired for the day — the drill was napping in his case and the hammer was locked in a long-term grudge with the linoleum.

But Stanley? He *buzzed to life*.

"This is it," he whispered, his LED blinking with heroic intensity. "This is my moment to tighten destiny itself."

He zoomed across the kitchen floor, found his trusty #2 Phillips head, and began his noble spin. *Whirrrr-ZZZZZzzz* went the screws. *Clack!* went the bolts. *Ka-chunk!* went something that may or may not have been the garbage disposal.

Bit by bit, the cabinet stood taller. Straighter. More proud. By the final screw, it was sturdier than ever — ready to face the next decade of canned goods and expired oatmeal.

The home owners cheered. The dog blinked approvingly. Even drill gave him a subtle nod from across the room.

And from that day forth, Stanley was no longer just "that jittery thing in the drawer."

He was *Stanley the Stabilizer* — Tightener of Hinges. Defender of Ikea Flat-Pack Furniture.

Phantom Load

A **Phantom Load** is something that is still draining juice, even when everything is turned off.

The toaster is asleep. The microwave simmers on standby. The fridge babbles softly about door seals and crisper drawers, "...the door is closed, is the light on or off?"

Eddie Current twitched. His instincts finely tuned by years of sizzling through socket suburbs and outlet alleyways — felt it. An electric drain. Subtle. Sinister. Constant.

He darted into a USB-C port and emerged tangled in a jungle of charging cords.

There they were:

~~ Smart Phones — millions of them all over the world, napping on nightstands, screen-locked but never truly off.

~~ iPads: — glowing faintly in sleep mode, dreaming of spreadsheets and cat videos.

Eddie narrowed his eyes.

"Phantom loads," he whispered. "Parasitic. Persistent. Pretty obnoxious."

"I knew it! I saw the voltage sag in Sector 9. Send backup — bring Charlie Capacitor. We're going to need stored energy and sass.

Just then, the wall trembled. Votyarella opened one smug headlight and yawned. "Who dares interrupt my 2 AM top-off cycle?"

Eddie didn't flinch.

"You've been overcharging, *Voltarella.* That kind of drain disrupts the harmony of the home. Step away from the socket."

The car growled — digitally.

"You're outdated, Current. I'm the future. Silent, electric, and ALWAYS online." said Volterra in sleepy *Phantom Load*

Before things could escalate into a full-blown power brawl, a blur of denim and diodes swung into view.

"Live with it — Live with us" they all shouted in unison. *"You don't have any choice or other option."*

Theater: Live & Electrified

Eddie Current once was only a humble Ambassador of Power — lighting bulbs, frying toast, and jumpstarting sleepy coffeemakers. But those days are long over.

Now Eddie also appears on stage!

Tonight's venue? The *Ampa-Theatre*, a chrome and neon coliseum where every performance is a 50-volt spectacle. The house lights dim. A hush falls over the audience. And then— **BOOM** — a thunderclap from twin 1400-watt woofers shakes the floorboards. Fog machines hiss. Luminous arcs dance across Tesla coils hidden in the wings. A voice crackles from the sound system:

> "He's amped. He's grounded. He's
> alternating at 60 cycles per second...
> He's **EDDIE CURRENT!"**

Eddie bursts onto the stage in a suit of LED-threaded mesh, each wire flickering with choreographed precision. His cape billows — thanks to a hidden bank of strategically placed oscillating fans — and his super tweeters let loose a high-frequency fanfare so crisp it makes your teeth buzz.

This isn't your grandma's stage production. This is *performance circuitry*.

Eddie Current Live and in Concert

As he sings his hit number, "Ride the Lightning and Call Me in the Morning," the graphic screens behind him flare into life. A giant wall of synchronized OLED panels depicts a psychedelic vision of electric pulses zipping through power lines, toaster wires, and disco balls.

The lighting design — controlled via DMX512 with Eddie's own biometric input — shifts from moody blue to apocalyptic red in time with the beat drop.

Stagehands, dressed as dancing surge protectors, pirouette across a grid themed floor that glows beneath their feet.

Each scene is hjeart pounding. When Eddie scales the peak of *Mount Voltage*, fog pours down like ionized mist. The soundtrack swells. Spotlights cross like saber beams. His boots crackle as he moonwalks across a glowing arc-furnace bridge, sparking cheers from the crowd.

By the finale, he raises his arms and the theater pulses with a bass note so deep, pacemakers file complaints.

The audience is left breathless, hair with static charge, faces glowing with ungrounded joy.

As the curtain falls, one thing is clear:

Eddie Current isn't just a performer anymore. *Entertainment is* the new grid *and Eddie is its star.*

ELECTRIC VEHICLES

The Power Team and Grid

- Automobiles
- Airplanes
- Bikes & Scooters
- Busses & Trucks

Behold **Voltarella,** the electric car with the soul of a ballerina and the zoom of an energy cocktail-laced falcon. Sleek as a dolphin in a tuxedo, she glides — not drives — down the boulevard, leaving behind only a gentle whiff of smug eco-superiority and maybe a single, spinning leaf which fell in her wake.

Imagine, if you will, a car so quiet it makes a library seem like a rock concert. So silent, it doesn't just sneak up on you — it *tiptoes* into your peripheral vision, offers a polite nod, and vanishes down the road before you can say, "Was that a...?"

Townsfolk never hear her coming. By the time someone turns to say, "Did you hear something — ?" *whoosh.* She's already a shimmering blur in the distance, making highway signs wobble in admiration.

Is she powered by starlight and whispered dreams? or a Fronk full of AA batteries? Mechanics whisper rumors of her regenerating brakes and her uncanny ability to find parking on the first try.

One day, she entered the annual "Vroom-Off" against roaring roadsters and fire-belching trucks. When the flag dropped, *Voltarella* vanished like a magician's trick card.

She was already sipping electrons at a charging station on the other side of town by the time the others hit second gear.

Another evening of happy conversation was just getting started at the Substation Saloon when **Wendy Watt Asked, "Tesla Was an AC Guy, But the Car Runs on DC?"**

The crowd leaned in as Breaker Bill wiped down the copper bar top and nodded sagely.

"Here's the thing," Eddie said, gesturing to the neon lightning bolt flickering above him. "People always ask: *'Wasn't Nikola Tesla the King of AC? So why does the Tesla car run on DC?'* And I say: Yep. And Yep again."

A few smart outlets blinked in confusion.

"Let's break it down," Eddie continued, tapping his mug. "Nikola Tesla was the champion of alternating current — AC. He fought the good fight so power could travel long distances without fizzling out like a lazy sparkler. AC was perfect for lighting up cities, powering factories, and giving Edison ulcers."

A few of the older electrons chuckled knowingly.

"But modern electric cars — like the sleek, silent, slightly smug Tesla auto — run on direct current. Why? Because batteries store DC, not AC. They're basically chemical warehouses of steady, one-directional flow. You don't want your electrons doing the cha-cha inside a battery. You want them marching forward like they mean it."

"Now," he added, swirling his electrolyte cocktail, "that doesn't mean Tesla the car betrayed Tesla the man. It just means the car converts AC to DC when charging — kind of like taking the freeway to the grocery store, then walking the last block with your groceries. Same destination. Different mode."

A buzz of understanding rippled through the saloon.

"So yes," Eddie concluded, "Nikola Tesla gave us AC for distance and scale. But when it comes to zipping around town in a whisper-quiet rocket sled on wheels? DC's got the wheel. Both are charged with the same electrons — just different dance styles."

Bob Transformer raised a glass. "To both Teslas — the dreamer and the driver!"

"Cheers!" shouted Andrea Amp, adjusting her copper earrings. "Now let's discuss dimmer switches, philosophically."

Electric Cars: The Zest of the West

Imagine a vehicle so silent, it sneaks up on you like a cat on tiptoes. That's an electric car for you — stealthy, sleek, and always ready to pounce on the pavement.

Plug and Play
Charging an electric car is as simple as plugging in your smartphone, but with a twist. Instead of scrolling through social media while it charges, you might find yourself pondering life's great mysteries, like why do we park in driveways and drive on parkways?

Range Anxiety: The New Frontier
Gone are the days of worrying about gas prices. Now, it's all about "range anxiety" — that thrilling game of "Will I make it to the next charging station, or will I become a modern day pioneer, forging a new path on foot?"

Eco-Friendly and Ego-Friendly
Driving an electric car isn't just good for the planet; it's great for your ego. Nothing says "I'm saving the world" quite like gliding past a gas station with a smug smile and a fully charged battery.

The Future is Electric
So, buckle up and enjoy the ride. With electric cars, every journey is a chance to experience the future — quietly, efficiently, and with a spark of humor.

And remember, in the world of electric vehicles, the only thing more charged than the batteries is the conversation.

Absolutely! Let's take a whimsical spin through the world of electric vehicles (EVs), where the roads are paved with puns, the journeys charged with humor.

Remember, in the world of electric taxis, the only thing more charged than the batteries is the conversation.

The Great Range Anxiety Rodeo

The **Voltarella** had shiny solar eyelids, regenerative brakes, and the cutest little charging port you ever did see, but she also had *range anxiety* — the deep, existential dread of running out of juice mid-journey, far from an outlet and surrounded only by judgmental gas stations.

Voltarella lived in the town of Ampville, where everyone else seemed blissfully confident in their battery levels. The big SUVs would thunder past, laughing exhaustedly, while Voltarella clutched her dashboard and muttered, "Seventy-three miles left... but what if the wind changes?

One day, she decided to face her fears and take the legendary Route 66 *unplugged* — a lonely scenic stretch with *no chargers for 48 miles*. That's right: no fast-charge stations, no friendly garage plugs, not even a single kind-hearted farmer with a solar panel.

As Voltarella rolled out of town, every bump in the road felt like a ticking time bomb of battery loss. Rabbits waved. A tumbleweed rolled by. Voltarella whispered to her navigation system, "Are you *sure* we have enough charge? What if we have to turn back?" At mile 20, a butterfly landed on the hood and she squealed, "That *must* cost at least 1% of charge!"

But just when panic peaked and Voltarella was calculating how many extension cords would reach from the last town, something magical happened.

A solar-powered smoothie truck named *Sunshine Sally* pulled alongside.

"You okay there, Miss?" she asked with a smile. "You're just running on stress and fumes!"

Voltarella nodded furiously. "I've got 31 miles left and the map says 30.8 to go. I could expire *emotionally*!" Sally smiled. "You've got this. Trust your electrons. Trust the road. Also — eat a banana. It helps."

With that, she zipped off, leaving behind good vibes and a faint scent of mango kale.

Voltarella tightened her metaphorical battery belt, turned up the optimism, turned down the air conditioner and coasted into the next town with *0.2 miles to spare*.

From that day on, she still worries, but just a little less. Because sometimes the road is long, but the charge, like courage, is just enough to pull her through.

The 80/20 Rule of Battery Zen...

Old Man Ohm

The sacred mantra whispered among battery sages, EV enthusiasts, and cautious phone users alike is...

Imagine your battery is a cranky old wizard named *Old Man Ohm*, who prefers moderate temperatures, early bedtimes, and *never* being too full or too empty. The idea behind the 80/20 rule is to keep *Old Man Ohm* happy — because when he's happy, your battery lives a long and powerful life.

Why not charge above 80%?

Because when you do, *Old Man Ohm* gets bloated. *"Too much power!"* he groans, holding his belly. "I can't hold all this energy without wear and tear on my innards!" At high charge levels, the lithium ions crowd in tightly, causing stress, overheating, and long-term grumpiness (a.k.a. capacity loss).

Why not drain below 20%?

Because *Old Man Ohm* gets hangry. *"You expect me to summon lightning on an empty stomach?"* he snaps. Low charge levels strain the battery's internal chemistry, force it to work harder and accelerate aging, especially if left low for too long.

So what should you do?

Charge to 80% for daily use. Avoid going below 20% unless you really have to. Go full 100% or down to 0% only for rare journeys or calibration — not daily life.

This way, you will be treating your battery like a temperamental but talented artist — feed it just enough, don't overwhelm it, and you'll both live harmoniously for many miles to come.

Electric Airplanes

The growth of electric aircraft is one of the most dynamic and transformative trends in aviation today. It is fueled by advancements in battery technology, environmental concerns, and demand for quieter, more efficient regional air transport.
Here's a breakdown of its development.

Fasten your lightning belt, and board the *Sky Wattage Express* — your electrified flight through the whimsical history and future of electric aircraft, narrated by none other than *Captain Mary Louise Watt* and her *Co-pilot Wendy Watt.* Enter the 2020's, when a jolt of progress struck like a lightning bolt! Sleek electric motors met snazzy lithium-ion batteries, and suddenly aircraft began zipping around like self-important hummingbirds with Masters diplomas.

Captain Watt herself took to the skies in her *Voltsvagon*, a lightweight electric trainer aircraft. She got her wings as rumors of a regional commuter were about to be released.

Rumors claimed it was designed as a strange breed of flying creature, called **eVTOLs**, *e*lectric *V*ertical *T*ake-Off and *L*anding craft.

These **eVTOLs** resembled oversized dragonflies crossed with soft drink machines. They promised to whisk commuters above traffic, latte in hand.

As battery wizards, motor images, and carbon-fiber cobblers joined forces, the newly formed Electric Sky Guild expanded its rank to include these new members.

* **Joby**, a hover-happy chariot maker

* **Archer**, who sculpted sky taxis with the grace of a screaming falcon

* **Airbus & Boeing**, two ancient titans now dabbling in hybrid spells

Meanwhile, the humans on the ground cheered! "Quieter skies! Cleaner air! Fewer engine goblins!"

Air Wars: The Coalition of Combustion

But not everyone was thrilled.

Lurking in the clouds was *The Coalition of Combustion* — a ragtag gang of jealous, old-school jet-liners with soot-stained wings and loud, gassy engines.

Their leader, *Captain Kerosene*, once flew solo from Detroit to Dubai and never stopped bragging about it.

"Back in my day," he growled, "planes had *guts*. We belched fire, we smelled like adventure, and *none of this recharging nonsense*!"

Determined to sabotage the electric uprising, Captain Kerosene and his sidekick *Petrolina* (a mothballed DC3 with a grudge and a leaky tank) planned to smog up the airspace with fake turbulence and fossil-fueled propaganda flyers.

In the epic showdown between Captain Kerosene and the rising fleet of Electric Planes, the skies crackled with tension — and innovation.

Are the skies big enough for both?

Captain Kerosene, trailing smoke and nostalgia, put up a roaring fight, but the quiet confidence of the Electric Squadron won the day with clean climbs, smooth glides, and zero tailpipe tantrums on short flights in densely populated cities.

As the propellers whirled on long flights with many passengers, and the batteries beamed, Captain Kerosene tipped his hat, conceding some routes, "You've got the spark, kid." The skies, once ruled by fumes, now buzz with the promise of a cleaner tomorrow.

In the bustling pastel-hued city of Voltville, sparks fly — not from overhead wires, but from sheer excitement and the whir of whisper-quiet wheels.

On the **Emerald E-bike of Destiny,** we find **Zip Zap the Courier,** cheeks flushed with wind and purpose, weaving through imaginary traffic cones of delight. His orange backpack bulges with mystery — perhaps burritos, perhaps letters of great romantic import. Who knows. Not even him.

But beside him — good heavens, she's gaining! — is **Granny Watt** (Wendy's grandmother), the scooter-riding whirlwind of wisdom. Donning her polka-dot headscarf and a grin that could short-circuit a streetlamp, she leans into her turbo-charged ride like she's late for a knitting tournament. Her orange scooter hums with pure gumption and just a hint of cinnamon.

Behind them, trees smile. Buildings lean in for a better look. Even the clouds above draw little lightning bolts in support.

This is not just a race — it's a revolution in motion. A spark-powered ballet of balance, bravery, and brrrrring-bells. And it's only the beginning. They've surged in popularity over the last decade, transforming urban transport, gig deliveries, and joyrides alike.

And standing proudly on the sideline, waving like the Mayor of Motion himself, is none other than William Watson, Electric Mobility Ambassador and inventor of the "Five-Finger Friendly Wave."

His green suit is starched, his glasses gleam with purpose, and his campaign button proudly reads:

"Watson Whirrrr!"

Yes, he's running for "Secretary of Scooters" in next week's public transit parade.

Behold! A new chapter rolls in on silent wheels,

Electric Busses!

Some are sleek, silent, and so smooth they seem to hover slightly above the road. They don't growl —they *glide*. They're powered not by gas, but by pure optimism and rechargeable batteries the size of a Thanksgiving table.

Others are the rolling whales of the asphalt sea — long, majestic, a little creaky, and full of mysterious humans staring out the window like wistful poets. They glide through cities and towns with an air of seasoned wisdom, puffing along with sighs of diesel or — lately — humming with the serene purr of electricity.

I'll never forget my school bus driver. His name was Earl, and he had a mustache shaped like a suspension bridge, and always chewed a cigar. He always had a story for us.

Today more school busses are going electric. They look the same, big, boxy and yellow. Our friend, Breaker Bill, with his contacts on the Power Grid, has an inside track for a part-time gig driving one of these school buses. He enjoys the time spent driving and telling Saloon stories to the students.

And best of all? No exhaust burps. Just a faint w*hoosh* and a cloud of smug satisfaction from Mother Earth.

They're the storytellers of the streets, gathering tales from passengers who missed their stop because they were reading a book, or carried plants to work like botanical bodyguards, or boarded just to say hello to their favorite driver.

In conclusion, busses — whether belching, buzzing, or bounding over speed bumps — are the unsung heroes of our public pathways. They gather us, move us, and quietly remind us that no one travels this road alone (especially during rush hour).

Amooz-Ing Delivery Van

Behold a zippy electric marvel that normally hums to your porch with the grace of a greased hovercraft. In the land of doorbell chimes and cardboard dreams, your wife peered out the window today expecting the arrival of Electra the delivery van powered by the green new deal itself.

But alas... no whirring wheels, no neatly-taped package, no slightly judgmental delivery photo of a box angled oddly agA.I.nst the hydrangeas.

What happened? Let us consider the suspects,

Forgetfulness Fog – A common glitch in the human operating system. Did you dream the order? Or merely add it to your cart and did it vanish like a vapor trail of intention?

Cart Abandonment Syndrome – A global epidemic where items sit in limbo, sobbing softly inside your digital shopping cart like puppies in an unadopted wish list.

Wrong Day Dilemma – Maybe the tracking email said "Tuesday" but your brain translated it as "NOW." Happens to the best of us.

Recharge Delay – Perhaps Electra is charging at a solar station, sipping electrons with her siblings. She'll arrive — eventually. Probably when you're in the shower.

Long-haul Trucker

Imagine a noble creature, part navigator, part snackologist, part audiobook philosopher, a road warrior wrapped in denim and dreams piloting a great mechanical beast named things like *Bertha*, *The Rolling Burrito*, or *Sir Gear-a-Lot.*

These trucks aren't ordinary vehicles — they're portable kingdoms. Within their cabs you'll find a shrine to coffee, a bed tucked behind the wheel like a nap-ready cave, and probably a rubber duck wearing sunglasses on the dash (a requirement under Trucker Law #72).

Each morning the long-haulers rise with the sun (or well before it), salute the day with a muffin the size of a hubcap, and fire up their mighty steeds.

The route stretches far from Cornfield Curve to Cactus Hollow, past Billboard Forests and Gas Station Villages with names like "Fuel 'n' Fly" and "Gator Snacks Plus."

And while the miles roll on, their soundtrack is a blend of Truckstop tales on CB radio. Deep podcasts about cryptids or cooking, and the soothing hum of 18 enchanted wheels whispering, "You're doing great."

Enter the new generation, Electric long-haul trucks, who glide across the land like battery-powered unicorns. Their names? Things like *Glidezilla*, *Teslor the Quiet*, or *Voltron Freighticus*.

They plug in at neon-lit charging havens, sip lightning smoothies, and leave behind only tire tracks and the scent of ambition.

They're quiet. They're clean. They're proud members of the "Wattage & Courage Guild."

Long-haul trucking is more than logistics — it's mythology on wheels.

They are the legend of humans and machines, criss-crossing a continent to keep everything moving — from cereal to saxophones.

So next time you see one barreling down the freeway, tip your hat or honk your thanks, because you just witnessed a hero of the highway, possibly named *Lugnut Larry*, steering a behemoth named *Mildred Thunderroll* through another chapter of the open-road epic.

Grid Crisis Caused by Data Centers

Eddie Current and the Cloud Crisis: Zapocalypse of the Zettabytes

The electric breeze was gentle that morning as Eddie Current zipped along the copper-lined clouds, sipping an Ohm smoothie and enjoying the quiet hum in the power grid. The ever-sparking Ambassador of Power entered the Substation Saloon and ordered zap-smoothie when the lights flickered — not the usual moody ambiance flicker, but a power-draining, grid-gulping kind of flicker.

What started with a flicker was just a hiccup in the grid. But Eddie knew something was off. The serenity didn't last. Alarms blared from his Pocket Ohmmeter, and his cape crackled with static. Something *big* was pulling juice — and fast.

Somewhere in the vast electric ecosystem, something was guzzling gigawatts like a racoon with an energy cocktail problem.

A blip on the Electro Radar confirmed it: DataZilla, the mega data center complex, had booted up a new AI fleet requiring enough power to toast every bagel on the East Coast simultaneously.

"This place is drawing more juice than the Saloon on karaoke night!" Eddie shouted, his hair crackling with concern.

Mary Louise Watt entered, piloting her "Efficiency Drone™, with **Bob Transformer** in tow, lugging his "Step Up / Step Down" briefcase.

"We're not just talking power surges anymore," Mary warned. "These AI companies are pulling petawatts, Eddie. If we don't manage the load, the Grid's going to fry like an overcooked strudel!"

Charlie Capacitor chimed in from a dusty corner with Capacitron™ flashing. "We need storage, fast. I'm buffering as hard as I can!"

Just then, Charlie, out of breath, said "Eddie! In addition, the *Data Cloud* is sucking up juice like a dehydrated supercomputer in a sauna!"

Eddie sprang up. "Not the *Data Cloud* again! I told them to throttle their wattage during peak meme-upload hours!"

Eddie frowned, sparks popping off his eyebrows. "Who just plugged into the Grid like it's an all-you-can-eat buffet? They're stressing out substations, waking up retired transformers, and even the electrons are filing complaints!"

Eddie dove straight toward the epicenter. Towering steel buildings blinked with activity. *Cloud Stor*age was puffing energy into cumulonimbus clouds. A.I. bots with glowing screens for faces paced impatiently, demanding more bandwidth and kilowatts like toddlers in a candy aisle.

The crisis had been building for months. *Cloud Storage,* an A.I. firm, multiplied data like binary bunnies, creating data centers the size of football stadiums, which were springing up everywhere — each humming, blinking, and slurping electrons like a neon octopus. *Cloud Storage* sounded fluffy, but Eddie knew it was anything but. It was more like thundercloud storage — heavy, stormy, and dangerously overcharged.

Bob Transformer Mary Louise Watt, and Eddie convened an *Emergency Grid Summit* on the spot. They huddled over a glowing schematic projected from Myron Volt's vintage voltmeter, now repurposed as a "Current Event Analyzer."

"We need a global strategy, one that balances our thirst for dog videos, digital assistants, and predictive sandwich ads — without frying the entire grid," Eddie declared.

Mary Louise Watt and Bob Transformer soon joined him at the scene. Mary unrolled a blueprint titled *Grid Sustainability & Sanity Plan, Rev. 4.0*, while Bob grunted and adjusted his Step Up / Step Down vest. Mary said "the A.I. load is no joke. One server farm in Nevada just enrolled in a gym because it needed cooling."

Bob Transformer adjusted his vest, "Step up, step down — we've got to get smarter, not just stronger."

So, Eddie and his team launched **"Project Low-Latency Lightning"** — a dazzling initiative that blended microgrids, quantum capacitors, and a league of super-smart electrons called The Packet Screamers™.

Their goal? To balance the delicate dance between electrons and algorithms — to keep the lights on, the A.I. polite, and the grid grooving without any brownouts at bedtime.

By the end of the week, AI was powered more responsibly, the data clouds floated a bit lighter, and Eddie was back to sipping his zap-latte. Until the next crisis, that is — when the industrial-grade quantum toaster uprising began.

At the core of it all stood **Captain Cloud**, a digitized wizard made entirely of binary code and bad metaphors. "We're training 18 billion parameters a second, Eddie!" Captain Cloud buzzed. "We need more power, more cooling, more EVERYTHING!"

Together, they proposed a plan:

Solutions to Data Drain Dilemmas rev 3.1

- Install solar fields on *Cloud Storage's* rooftops.
- AI-driven power throttling — ironically, managed by AI to limit AI.
- Create Nighttim*e Grid Napping,* a policy requiring non-critical calculations to take power naps when human eyes are closed.
- Reroute excess charge to night-shift wind farms.
- Recalibrate the voltage to accommodate AI peaks during peak pizza-ordering hours.

Eddie raised his hand, the air buzzing. "Power isn't infinite, folks. If AI wants to grow, it has to grow smart. Let's electrify responsibly."

With a zap of agreement, the AI bots nodded (some simulated their nods) and Captain Cloud agreed to throttle his thunder.

Eddie rallied the neighborhood of electrons like a tiny blue union of spark-workers. "This is our turf, team! Let's balance these bytes!"

That night, Eddie flew home under a slightly dimmer sky — less bright, but more balanced. His cape flapped proudly as he smiled. "Team! We balanced those bytes. You all were part of the solution!" Eddie wiped his brow. "Power to the people — and a little less to the people's cloud."

The neon Cloud groaned and shrank to a polite mist.

And with that, balance was restored — at least until the next software update.

Power Grid Elders Debate The Great Tariff** Tussle

(An Accurate Review of Electric Rates)

Back at the Substation Saloon, the conversations turned to tariffs. After a long pause, Breaker Bill filled another stein with Lumen-Aid and continued with an earlier question about Electric Rates.

"Are you still arguing tariffs? Also a hot topic in the social media."

Myron said, "I say we bill'em by how emotionally attached they are to their devices. You want to plug in six smart speakers and a digital bonsai tree?

That's a *luxury load*, my friend. And should be in Tier 6, whereas all Smart Phones and GPS devices are critical and must remain in Tier 1 with the lowest rate."

Charlie jumped in with "Heating pads are both a necessity and a luxury, so put them in middle, Tier 3."

Andrea shouted "Hair dryers must remain with toasters in Tier 1."

Everyone nodded slowly. Eddie scribbled something on his clipboard: He now had another mission.

Tariff Proposal #7: Emotional Load Pricing + Toaster Equity Fund.

He jumped up. "Let's take it to the regulators."

**The terms "rate structures" and "tariffs" are often used interchangeably in the context of utility companies. While "rates" may specify the price per unit of consumption, "tariffs" provide the complete pricing framework and associated conditions. Also electric rates are a part of the overall tariff structure. Rates are specific dollar amounts per unit of consumption (like per kWh of electricity), while tariffs encompass a broader framework that includes these rates, as well as additional fees and conditions.

Watts the Deal? Here is a review of the Grid's Most Dramatic Divas and Budget-Friendly Darlings. The grid is interconnected from coast to coast. Why such a variance in rate (tariff) structures when the electric cost is virtually the same state to state?

Highest Rate: **CALIFORNIA (31.66¢/kWh)**
Advertising phrases:
* Where else are you going to go, darling?
* YOU need it, WE got it!
* We Invest heavily in grid fashion (wildfire safety upgrades)
* Obsessed with renewables — we compost our own voltage
* Solar subsidy for the wealthy, but... everyone else foots the bill

Lowest Rate: **TENNESSEE (13.04¢/kWh)**
Advertising phrases:
* My toaster did not have to take out a second mortgage to toast breakfast.
* Country charm with hydropower legs
* I light up hearts — on a budget
* Love hydroelectric and nuclear — clean, classic, efficient
* Bake biscuits and balance the load at the same time

Honorable Mentions:

New York (24.37¢/kWh): *"Broadway Voltage"* — dramatic, ambitious, wants to save the planet and also star in *Hamilton*.

Vermont (22.29¢/kWh): *"Green Grid Granny"* — lives off hydro, composts electrons, and knits with fiber optic cables.

Texas (15.32¢/kWh): *"Grid Cowboy"* — unpredictable but charming, shows up with wind and gas, sometimes forgets to winterize.

Florida (14.20¢/kWh): *"Sunshine Surfer"* — relaxed, stable, solar-kissed, and usually on time... unless there's a hurricane party.

Let's crank the voltage of absurdity up a notch and take a stroll through:

The Absurd Effects of Energy Tariffs:
A User's Guide To Rates/Tariffs
"Because sometimes your toaster pays more in taxes than you do."

The Toast Tax Panic
With time-of-use tariffs in full swing, toast lovers across America are waking at 3:47 AM just to get that sweet, low-voltage pricing window. Entire neighborhoods now glow softly before dawn, as bagels are sacrificed under the flickering judgment of dynamic pricing.

"Mom, why do we only eat breakfast in the dark now?"

"Because son, the grid is moody and peak pricing is cruel."

Washer Shame
High tariffs have turned laundry into a public performance art. Neighbors now passive-aggressively audit one another's spin cycles, and washing socks at 5 p.m. is grounds for exile in some HOA covenants.

"Sorry Karen, this cul-de-sac runs on off-peak tumble dry. You can take your lint drama elsewhere."

Solar's Existential Crisis
Solar panels, once the golden children of the grid, now lie awake wondering if net metering is just a polite lie. They give all day, basking gloriously, only to be told their electrons are *worth less* during their big performance hours.

"I used to be the future," whispered a solar array. "Now I'm just a glorified patio shade."

"You can plug in," said the EV app, "but it's going to cost you 38 cents and your last shred of optimism."

Rate Guilt

Light Bulb Guilt Therapy
and the great LED conversion. (This could be the next book)

EV Charging Roulette
Electric vehicle owners now resemble gremlins with spreadsheets. They whisper to their dashboards, sacrifice USB cables to appease the utility gods, and fight strangers at Walmart chargers over who remembered to check real-time pricing first.

Rate Guilt

Symptoms include apologizing to your light switches, excessive dimmer fondling, and crying softly when the thermostat clicks on.

> *"Every time I flip a light on, or walk out of a room with the light still on, I imagine Eddie Current shaking his head in disappointment..."*

Appliance Rebellion
Smart fridges now unionize. Microwaves blink passive-aggressively when summoned during peak hours. One brave toaster has already published a memoir: *"Crumbs of Resistance: How I Refused to Toast at 5 PM"*

Closing Thought:

Tariffs were meant to guide usage...
Instead, they've created a nation of night owls, spreadsheet monks, and lightbulb therapists. The grid isn't just alive — it's judging you.

"You can't spell *TARIFF* without *RIFF*, and baby, this grid is jazzing on anxiety."

Executive Conclusion: *Tariff Proposal #7*

With rising demand, time-of-use pricing confusion, and smart appliances developing sentience, the electric grid must evolve its pricing to reflect not just usage, but intention, emotional attachment, and circuit dignity.

This summary outlines the **E.L.P. (Emotional Load Pricing)** model, supported by a diverse council of electrical pioneers who barely agree on voltage polarity but stand united for fairer tariffs.

Core Policy Proposals:

The Emotional Load Multiplier Devices rated not just by wattage, but by **how much people *love* them.** example:
Refrigerator = Basic Rate Vintage
Lava Lamp = Sentimental Surcharge Tier 4

Kombucha-heating Wi-Fi-Enabled Mug Warmer = Luxury Load Tier 3

Toaster Equity Fund (TEF) redistributes grid
access subsidies to underrepresented appliances (e.g., analog coffee makers, unplugged George Foreman grills).

Led by Mary Louise Watt's campaign: *"Toast Is a Right, Not a Privilege."*

Time-of-Use Reimagined, Circadian-Circuit Pricing Introduced, synced with planetary mood swings and residential napping habits.

Bonus: "Nap Hour Discount" from 2–3 p.m.
Penalty: "Guilty Streaming Surcharge" from 1–4 AM

Appliance Based Voting Rights

Connected devices earn one vote per kilowatt-hour consumed, but smart toasters are limited to advisory roles due to recent burnt bagel scandals.

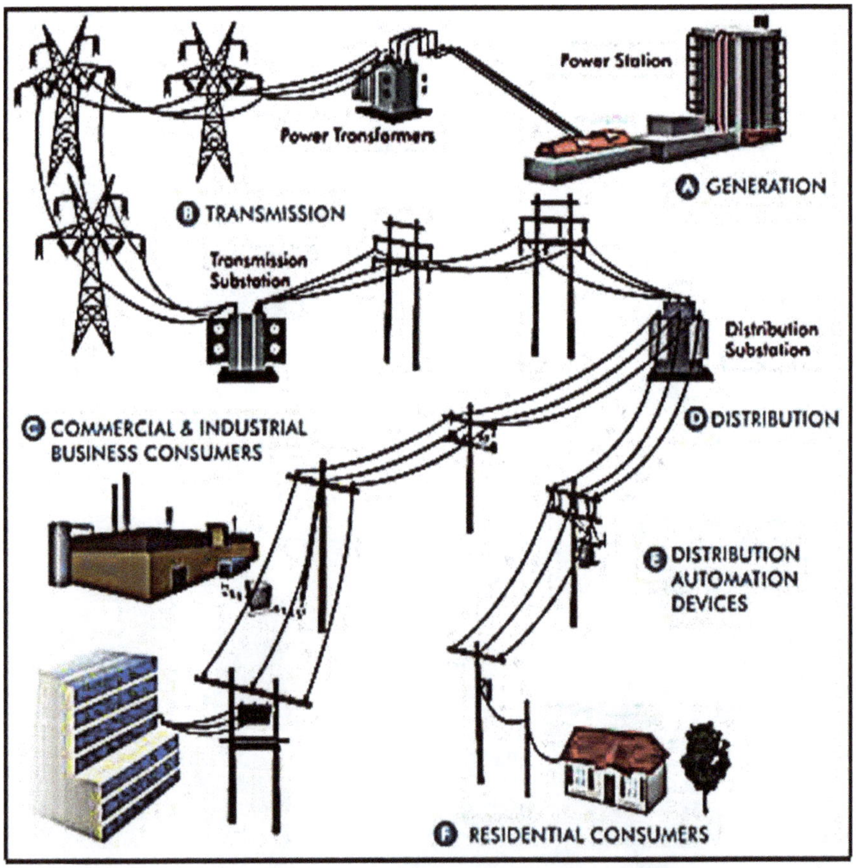

How many points can you identify where government agencies can Assess Tariffs on electricity?

And sooo, Dear Friends, Associates, Lighting Designers, and all those countless professionals with an electrical background, we have reached the end of the line. But don't worry... it's not the end of Eddie Current and his friends because electricity never truly stops it just finds another circuit to flow through, in perpetuity.

Long after we are all lost in the mists of time, Eddie and his friends will be cavorting in the circuits and charging on to new adventures.

Whether you're flipping a switch, charging a gadget, or explaining the Theory of Electricity to your toaster, your brother, or a friend, remember this: *the power is inside you all along.*

Stay grounded, stay curious, and if you ever feel lost — just follow the hum.

~~ *Eddie Current, signing off* ~~

End of Line (Until Recharged)

~~~ A Final Thought from Eddie ~~~

"Understanding electricity isn't about memorizing numbers, it's about knowing how things behave when you flip the switch."

Thanks for bringing Eddie Current into your classroom, shop, or training pace. He promises to ask good questions and stay away from live panels.

Note: ***This book is designed to support conceptual understanding and discussion. It is not a substitute for hands-on electrical training or safety instruction***.

Educator & STEM Leader Guide

How to Use This Book in Classrooms, Libraries, and Learning Spaces

Hello, Educators! I'm **Eddie Current**, your friendly *Ambassador of Power*. My job is to keep electricity moving and help curious minds understand how it works without blowing a fuse.

This book was created to support conceptional understanding of electricity and **STEM learning through storytelling, humor**, and **real-world examples.** While Eddie and friends do the entertaining, the science underneath is an ideal companion for **high school STEM courses, trade programs, and workforce training environments**.

Intended Audience & Grade Range

This collection is well suited for:

~~ **Middle School (Grades 6–8)** – Concept reinforcement & discussion

~~ **High School (Grades 9–12)** – Concept review, literacy crossover

* Introductory electricity concepts
* Reinforcement alongside math based courses
* STEM literacy and career exploration

~~ **Trades & Technical Schools**

* Electrical, HVAC, construction, and industrial programs
* Pre-apprenticeship and workforce training
* Safety-focused conceptual understanding

~~ **Adult Learners & Career Changers**

* Refresher learning, foundational understanding before hands-on training

Core STEM Concepts Covered

Students are introduced to foundational electrical ideas, including:

~~ **Electric current** – how electricity flows

~~ **Voltage** – what pushes electricity along

~~ **Resistance** – why electricity sometimes slows down

~~ **Circuits** – closed paths that allow energy to move

~~ **Electric Utilities** — and rate structures

~~ **Energy generation** – where electricity comes from

~~ **Real-world applications** – homes, devices, transportation

Classroom & STEM Activities (Safe & Simple)

~~ **Electricity Spotting**

* Ask students to list everything in the room that uses electricity.
* Group items by *purpose* (light, heat, motion, sound).

~~ **Circuit Storytelling**

* Have students draw a simple "story map" of electricity traveling from

— A power source to a device—using characters instead of symbols.

~~ **Resistance in Real Life**

* Discuss examples of resistance outside electricity (traffic, water flow, teamwork). Connect the idea back to circuits.

~~ **Vocabulary Match-Up**

* Create cards with terms (current, voltage, resistance) and definitions.
* Let students match them—or invent Eddie-style explanations.

~~ **Design a Wired Friend**

* Students design their own "electrical character" and explain what role it plays in a circuit.

(All activities are discussion and creativity-based. No wiring required.)

Discussion Questions (Secondary & Adult Learners)

Use these prompts to support discussion, reflection, or written responses:

~~ **Why is electricity** difficult to visualize, and how does storytelling help?

~~ **How does thinking** of electricity as a "system" improve understanding?

~~ **How do Eddie's metaphors** compare to what you see in real equipment?

~~ **Why is conceptual understanding** important *before* working hands-on?

~~ **Where do you encounter** electrical systems in your daily life or trade?

~~ **How does power generation** affect how electricity is delivered and used?

Mini Glossary (Trade-Friendly Language)

~~ **Electricity** — energy that moves and performs work
~~ **Current** — the flow of electrical energy
~~ **Voltage** — the force that drives current
~~ **Circuit** — a complete path for electricity
~~ **Power Source** — where energy begins

Best use

~~ A **conceptual companion** to lab or shop work
~~ A **pre-lab discussion tool**
~~ A **Low-barrier introduction** to electrical thinking

(No formulas required. Eddie explains the "why" so instructors can teach the "how.")

Curriculum & Program Alignment

This collection supports:

~~ **NGSS concepts** (energy, systems, cause & effect)
~~ **Career & Technical Education (CTE)** pathways
~~ **STEM literacy and workforce readiness**
~~ **ELA crossover** (reading comprehension, metaphor, explanation)

Administrative Summary

The Curious Mind's Complete Theory of Electricity provides:

~~ A low-barrier entry into electrical thinking
~~ Improved comprehension before hands-on work
~~ Increased learner engagement and retention
~~ Support for diverse learning styles
~~ A professional, safety-conscious instructional tone

Note: This book is designed to support conceptual understanding and discussion. It is not a substitute for hands-on electrical training or safety instruction.

Endorsement for Tesla To Toasters To Teslas:

What stands out immediately is that this book isn't trying to teach electricity the way textbooks do. It's trying to help readers **finally make sense of it**.

Instead of starting with formulas or abstractions, the book starts with curiosity, why things work, where power actually goes, how ideas evolved, and why modern conversations around energy, transportation, and climate often collapse under their own confusion. The humor isn't decorative; it's structural. It lowers resistance so readers stay with concepts that would otherwise feel inaccessible or intimidating.

The book is distinct in its willingness to live in the middle ground between education and provocation. Not just explaining AC vs. DC or historical figures like Tesla and Edison; you're poking at the absurdities of how we talk about energy today, corporate messaging, public misunderstanding, political oversimplification, while still grounding everything in factual reference and visual support.

Books like this don't struggle because they lack content. They struggle because **they don't fit neatly into a single shelf**.

This is not a children's science books. It's not academic engineering text. It's not pure satire. It appeal instead to readers who are intellectually curious, system-oriented, and tired of being talked down to, people who enjoy learning when it's presented with wit, skepticism, and real-world relevance

Kate Stevenson

Moderator, UK Book Club on Goodreads

Kate - Hampshire, UK, The United Kingdom (898 books) | Goodreads

www.ingramcontent.com/pod-product-compliance
Lightning Source LLC
LaVergne TN
LVHW070851210926
848482LV00016B/913